STUDENT'S SOLUTIONS MANUAL

MILLER & FREUND'S PROBABILITY AND STATISTICS FOR ENGINEERS

EIGHTH EDITION

Richard A. Johnson

University of Wisconsin—Madison

Prentice Hall
is an imprint of
PEARSON

The author and publisher of this book have used their best efforts in preparing this book. These efforts include the development, research, and testing of the theories and programs to determine their effectiveness. The author and publisher make no warranty of any kind, expressed or implied, with regard to these programs or the documentation contained in this book. The author and publisher shall not be liable in any event for incidental or consequential damages in connection with, or arising out of, the furnishing, performance, or use of these programs.

Reproduced by Pearson Prentice Hall from electronic files supplied by the author.

Publishing as Prentice Hall, 75 Arlington Street, Boston, MA 02116

ISBN-13: 978-0-321-64169-4
ISBN-10: 0-321-64169-8

Prentice Hall
is an imprint of

PEARSON

www.pearsonhighered.com

CONTENTS

PREFACE

This students' manual is intended to help the student gain an improved understanding of the subject by providing model solutions for all of the odd numbered exercises in the text. The chapters in this manual correspond to those in the text. In many problems, we have broken the calculations down into steps of partial results to help the student better understand the calculations. However, the student should be aware that this does lead to more roundoff error than doing the complete calculation on the computer or calculator.

In the spirit of quality improvement, we would appreciate receiving your comments, corrections and suggestions for improvements.

Richard A. Johnson

Chapter 1

INTRODUCTION

1.1 The statistical *population* could be the collection of air quality values for all U. S. based flights flown during the period of the study. It could also be expanded to include all flights for the year or even all those that could have conceivably been flown. The *sample* consists of the measurements from the 158 actual flights on which air quality was measured.

1.3 a) A laptop owned by student.

b) Weight of laptop.

c) Collection of numbers, for all student owned laptops, specifying the weight of the laptop.

1.5 A hard drive is the population unit and the distance between the head and the disk is the variable. The statistical population is the collection of distances for all drives that are manufactured. It could, more abstractly, be the collection of distances for all those manufactured or could be manufactured. The sample consists of the collection of distances that are measured.

1.7 a) For the new sample, $\overline{x} = 214.67$ and the chart is shown in Figure 1.1.

b) The new $\overline{x}$ is below the lower control limit LCL $= 215$ so the process is not yet stable.

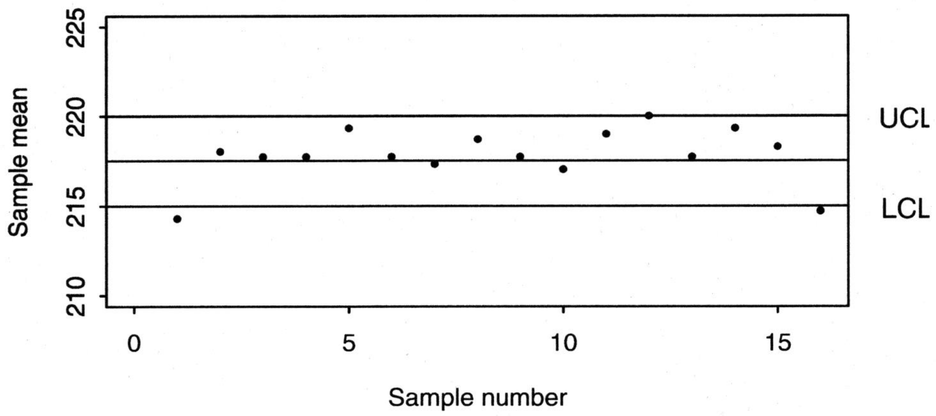

Figure 1.1: Xbar chart for Exercise 1.7

Chapter 2

ORGANIZATION AND DESCRIPTION OF DATA

2.1 Pareto chart of the accident data

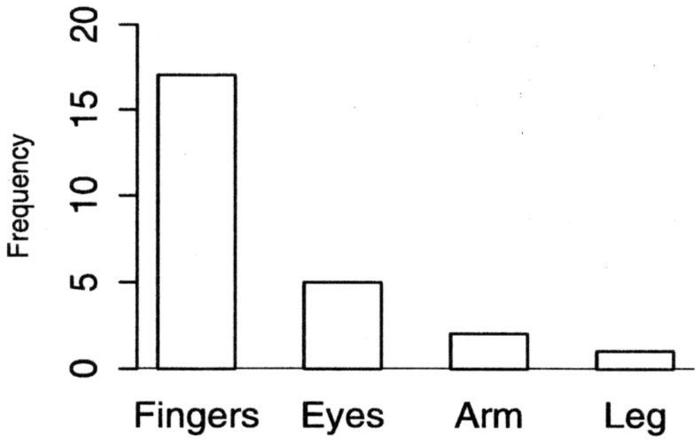

2.3 The dot diagram of the energy is

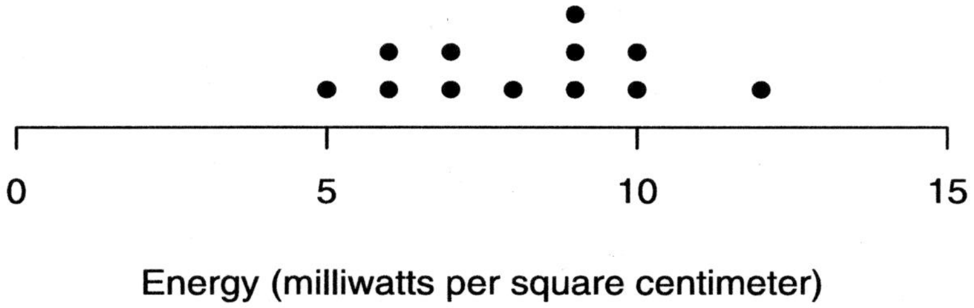

2.5 The dot diagram of the suspended solids data reveals that one reading, 65 ppm, is very large. Other readings taken at about the same time, but not given here, confirm that the water quality was bad at that time. That is, 65 is a reliable number for that day.

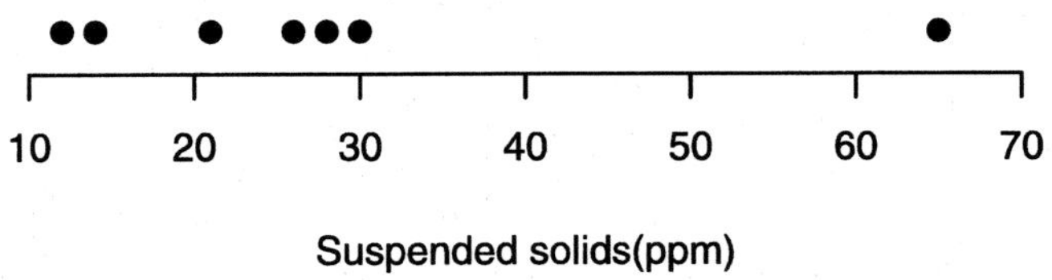

2.7 a) The class marks are 150.0, 170.0, 190.0, 210.0, 230.0 and 250.0.

b) The class interval is 20.

2.9 a) $20.5 = (16 + 25)/2$, 29.5, 38.5, 47.5, 56.5, 65.5 and the first class boundary is $16 - 4.5 = 11.5$

b) The common class interval is $9 = 20.5 - 11.5$.

2.11 We first convert the data in the preceeding exercise to a "less than or equal" distribution. The ogive is plotted below.

x	No. less than or equal
60.0	0
70.0	5
80.0	16
90.0	25
100.0	43
110.0	49
120.0	50

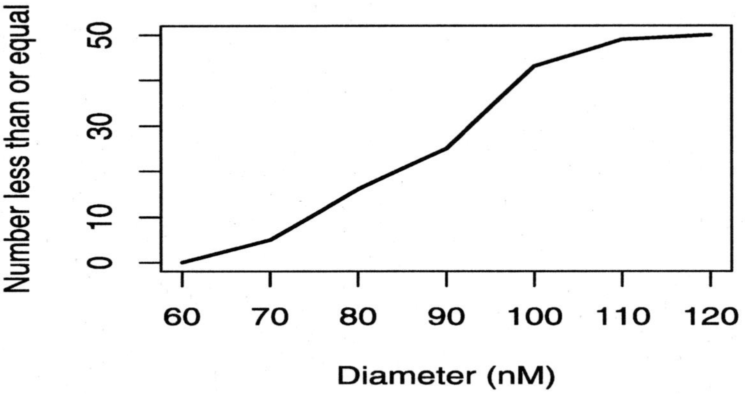

2.13 We first converting the data in the preceeding exercise to a "less than" distribution. The ogive is plotted below.

x	No. less than	x	No. less than
1.0	0	8.0	68
2.0	10	9.0	72
3.0	20	10.0	76
4.0	29	11.0	77
5.0	40	12.0	79
6.0	52	13.0	80
7.0	62		

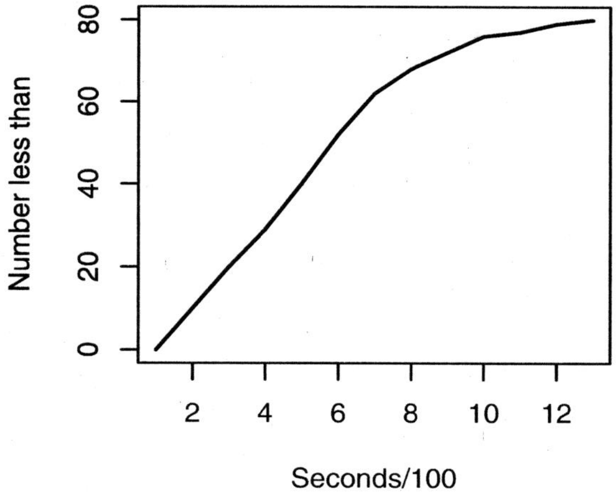

2.15 The "less than" distribution of the data in the preceeding exercise is:

Class boundary	Number less than	Class boundary	Number less than
20.0	0	60.0	60
30.0	4	70.0	80
40.0	17	80.0	94
50.0	35	90.0	100

The ogive is

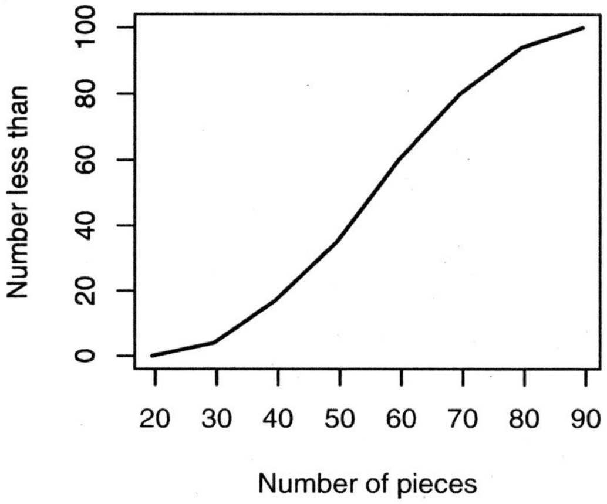

2.17 The cumulative "or more" distribution is:

Class boundary	Number equal or more	Class boundary	Number equal or more
-0.5	60	3.5	15
0.5	45	4.5	7
1.5	33	5.5	2
2.5	22	6.5	0

The ogive is

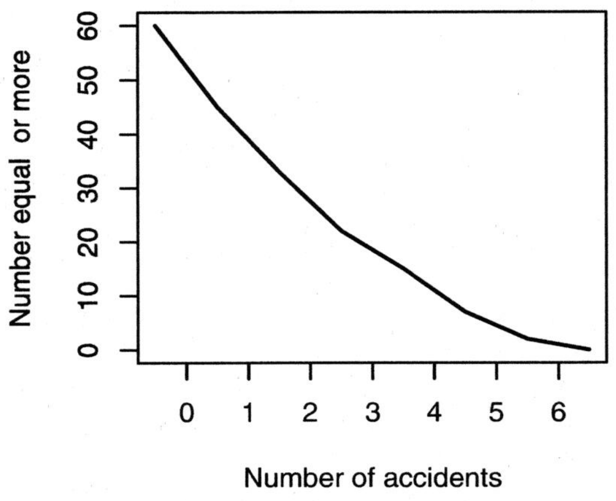

2.19 No. We tend to compare areas visually. The area of the large sack is far more than double the area of the small sack. The large sack should be modified so that its area is slightly more than double that of the small sack.

2.21 The empirical cumulative distribution function is

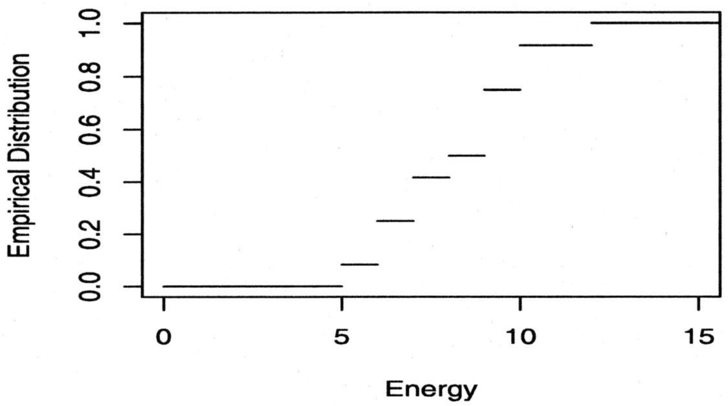

2.23 The stem and leaf display is:

1**	
2**	67, 88, 95
3**	17, 55, 70, 83, 91
4**	05, 19, 34, 62
5**	08, 40
6**	12

2.25 The stem-and-leaf display is:

2*	1 2
2·	6 8
3*	2 3 4 4
3·	5 5 5 5 6 6 7 8 9
4*	0 0 1 1 2 3 3 4
4·	5 5 5 5 6 7 7 8 8 9
5*	0 0 0 1 1 1 2 2 2 3 3 3 4 4
5·	5 5 5 6 6 6 7 7 8 9 9
6*	0 0 0 1 1 1 2 2 2 3 3 4
6·	5 5 5 7 7 8 8 8 9
7*	0 0 2 3 3 4 4 4
7·	5 6 6 7 8 9
8*	0 2 2 4
8·	5 8

2.27 (a) Would like a high salary-outlier that is large.

(b) Would like a high score-outlier that is large.

(c) Near average.

2.29 Greater on the mean. It would not influence the median for sample size 3 or more.

2.31 (a) $\overline{x} = (-6+1-4-3)/4 = -3$

(b)

$$s^2 = \frac{(-6-(-3))^2 + (1-(-3))^2 + (-4-(-3))^2 + (-3-(-3))^2}{4-1}$$

$$= (9+16+1+0)/3 = 8.667 \quad \text{so} \quad s = \sqrt{8.667} = 2.94.$$

(c) The mean of the observation minus specification is -3 so, on average, the observations are below the specification. On average, the holes are too small.

2.33 (a) $\overline{x} = (35+37+38+34+30+24+13)/7 = 211/7 = 30.14$ mm/min.

(b) The mean does not provide a good summary. The most an important feature is the drop in perfomance as illustrated in time order plot of distance versus sample number. It is likely the tip of the leg became covered with debris again. The early trials indicate what can be expected if this problem can be avoided.

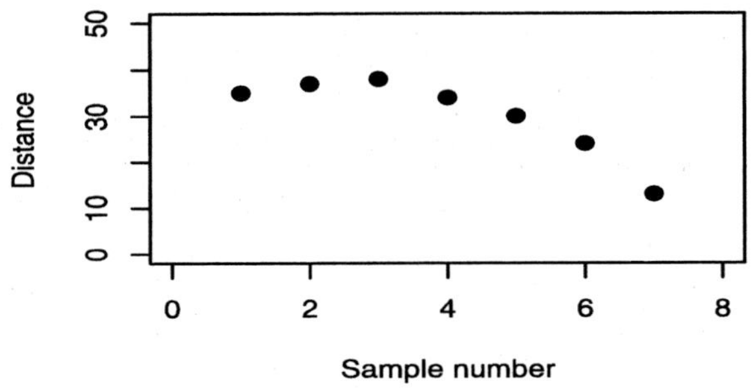

2.35 No. The sum of the salaries must be equal to 3 (175,000) = 525,000 which is less than 550,000. This assumes that all salaries are non-negative. It is certainly possible if negative salaries are allowed.

2.37 (a) The mean agreement is:

$$\frac{0.50+0.40+0.04+0.45+0.65+0.40+0.20+0.30+0.60+0.45}{10}=\frac{3.99}{10}=0.399$$

(b)

$$s^2=\frac{n\sum x_i^2-(\sum x_i)^2}{n(n-1)}=\frac{10(1.8891)-(3.99)^2}{10\times 9}=0.0330$$

Consequently, $s=\sqrt{0.0330}=0.1817$.

2.39 (a) The mean is 8.

(b) The sorted data are:

1, 2, 2, 3, 5, 6, 8, 9, 9, 10, 10, 10, 13, 15, 17.

The median is the eighth smallest which is 9.

(c) The boxplot is

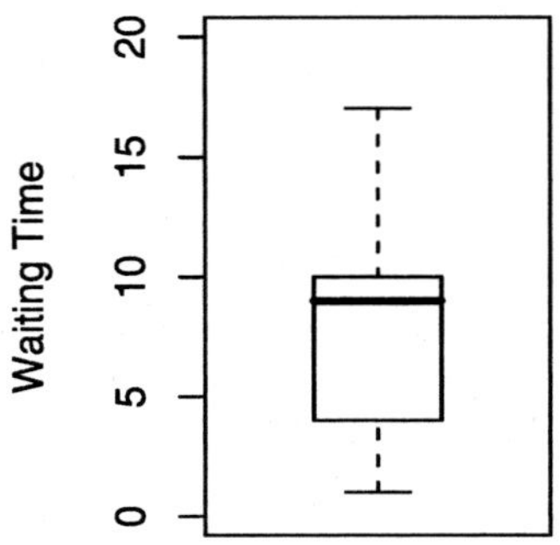

2.41 The mean is 30.91. The sorted data are:

29.6, 30.3, 30.4, 30.5, 30.7, 31.0, 31.2, 31.2, 32.0, 32.2

Since the number of observations is 10 (an even number), the median is the average of the 5'th and 6'th observations or $(30.7 + 31.0)/2 = 30.85$. The first quartile is the third observation, 30.4, and the third quartile is the eighth observation, 31.2.

2.43 The box plot is

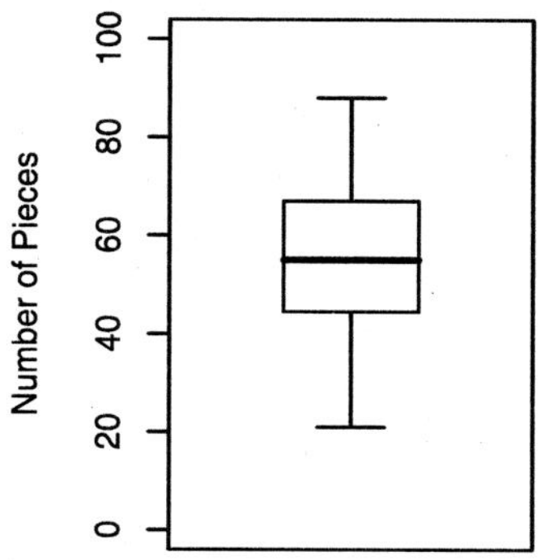

2.45 (a) For the non-leakers $\overline{x} = (.207 + .124 + .062 + .301 + .186 + .124)/6 = .16733$

(b) Also, $.207^2 + .124^2 + .062^2 + .301^2 + .186^2 + .124^2 = .20264$ so

$$s^2 = \frac{6(.20264) - (1.004)^2}{6 \cdot 5} = .006928 \qquad \text{so} \qquad s = \sqrt{.006928} = .0832$$

(c) The means are quite close to each other and so are the standard deviations. The size of gap does not seem to be connected to the existence of leaks.

2.47 The class marks and frequencies are:

Class mark	Frequency
65.0	5
75.0	11
85.0	9
95.0	18
105.0	6
115.0	1

Thus, $\sum x_i f_i = 4370$, $\sum x_i^2 f_i = 389,850$. The mean is $4370/50 = 87.40$, and $s^2 = (389,850 - 4370^2/50)/(49) = 161.47$.

2.49 The class marks and frequencies are:

Class mark	Frequency
24.5	4
34.5	13
44.5	18
54.5	25
64.5	20
74.5	14
84.5	6

We first find $\sum x_i f_i = 5550$ and $\sum x_i^2 f_i = 331525$. Further, $\bar{x} = 5550/100 = 55.5$, and $s^2 = (100 \cdot 331525 - 5550^2)/(100 \cdot 99) = 237.37$ so $s = 15.407$. Consequently, $v = 100s/\bar{x} = 27.76$ percent.

2.51

$$\left(\sum_{i=1}^{n}(x_i - \bar{x})^2\right)/(n-1) = \sum_{i=1}^{n}(x_i^2 - 2x_i\bar{x} + \bar{x}^2)/(n-1)$$

$$= \left(\sum_{i=1}^{n} x_i^2\right)/(n-1) - \left(2\bar{x}\sum_{i=1}^{n} x_i\right)/(n-1) + n\bar{x}^2/(n-1)$$

$$= \left(\sum_{i=1}^{n} x_i^2\right)/(n-1) - 2\left(\sum_{i=1}^{n} x_i\right)^2/(n(n-1)) + \left(\sum_{i=1}^{n} x_i\right)^2/(n(n-1))$$

$$= \left(\sum_{i=1}^{n} x_i^2\right)/(n-1) - \left(\sum_{i=1}^{n} x_i\right)^2/(n(n-1)) = \left(n\sum_{i=1}^{n} x_i^2 - \left(\sum_{i=1}^{n} x_i\right)^2\right)/(n(n-1))$$

2.53 (a) The median is the 50/2 th = the 25th largest observation. This occurs at the right hand boundary so the median is 90. Applying the rule,

$$90 + \frac{0}{18} \times 10 = 90$$

(b) There are 80 observations. There are 40 observations in the first four classes, and 40 in the last four. Thus, the estimate for the median is the class boundary between the 4'th and 5'th class. We could use 5.0.

2.55 (a) There are 50 observations. Q_1 is in the second class and Q_3 is in the 4th class. The lower class boundary for the 2nd class is 245. There are 3 observations in the first class and 11 in the 2nd class. The class interval is 40. Thus, Q_1 is estimated by $40(50/4 - 3)/11 + 245 = 279.55$ Proceeding in a similar fashion gives $325 + 40(37.5 - 37)/9 = 327.22$ as the estimate of Q_3. Thus, the interquartile range is 47.67.

(b) There are 80 observations and $80/4 = 20$ which is on the boundary of the second and third classes so the estimate for Q_1 is 3.00 and for Q_3 it is $1((60 - 52)/10 + 6.00 = 6.80$.

2.57 (a) The weighted average for the student is

$$(69 + 75 + 56 + 72 + 4 \cdot 78)/8 = 73.0.$$

(b) The combined percent increase for the average salaried worker is:

$$(24 \cdot 60 + 33 \cdot 30 + 15 \cdot 40)/(24 + 33 + 15) = 42.08 \text{ percent}.$$

2.59 (a) From the ordered data, the first quartile $Q_1 = 1,712$, the median $Q_2 = 1,863$ and the third quartile $Q_3 = 2,061$.

(b) The histogram is

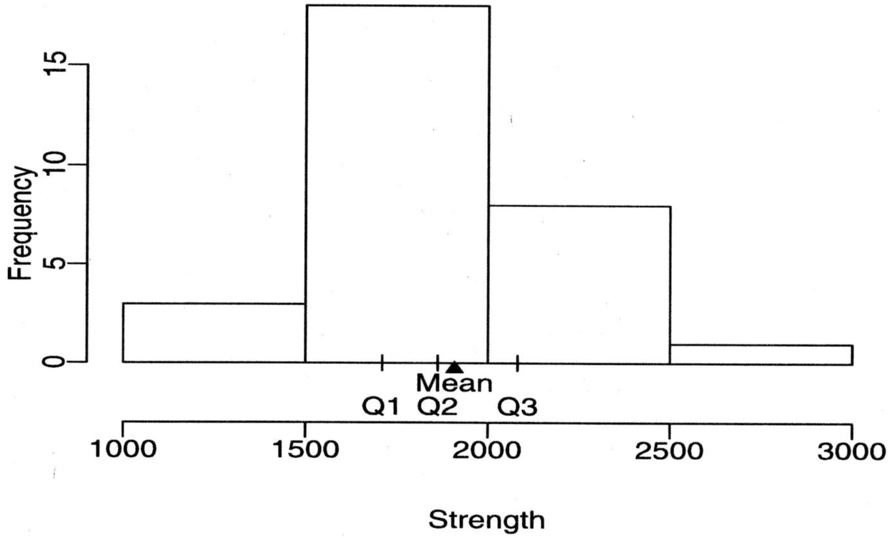

(c) For the aluminum data, we first sort the data.

66.4	67.7	68.0	68.0	68.3	68.4	68.6	68.8
68.9	69.0	69.1	69.2	69.3	69.3	69.5	69.5
69.6	69.7	69.8	69.8	69.9	70.0	70.0	70.1
70.2	70.3	70.3	70.4	70.5	70.6	70.6	70.8
70.9	71.0	71.1	71.2	71.3	71.3	71.5	71.6
71.6	71.7	71.8	71.8	71.9	72.1	72.2	72.3
72.4	72.6	72.7	72.9	73.1	73.3	73.5	74.2
74.5	75.3						

The first quartile $Q_1 = 69.5$, the median $Q_2 = 70.55$ and the third quartile $Q_3 = 71.80$. The histogram is

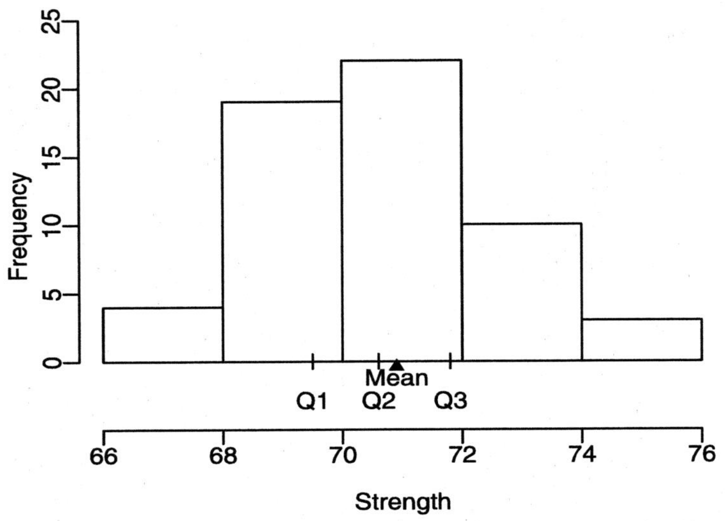

2.61 (a) The frequency table of the aluminum alloy strength data (righthand endpoint excluded) is

Class limits	Frequency
[66.0, 67.5)	1
[67.5, 69.0)	8
[69.0, 70.5)	19
[70.5, 72.0)	17
[72.0, 73.5)	9
[73.5, 75.9)	3
[75.0, 76.5)	1

(b) The histogram, using the frequency table in part (a), is

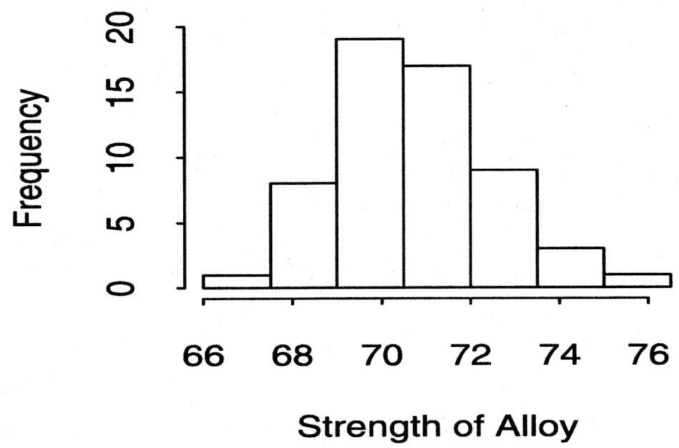

2.63 (a) The mean and standard deviation for the earth's density data are

$\bar{x} = 5.4835$ and $s = 0.19042$

(b) The ordered data are

5.10 5.27 5.29 5.29 5.30 5.34 5.34 5.36 5.39 5.42 5.44 5.46

5.47 5.53 5.57 5.58 5.62 5.63 5.65 5.68 5.75 5.79 5.85

There are $n = 23$ observations. The median is the middle value, or 5.46. The must be at least 23/4=5.75 observations at or below Q_1 so Q_1 is the 6th larges value. $Q_1 = 5.34$. condition. Similarly, there must be at least 3(23/4)=17.25 observations at or below Q_3 so Q_3 is the 18th largest observation. $Q_3 = 5.63$.

(c) From the plot of the observations versus time order we see that there is no obvious trend although there is some suggestion of an increase over the last half of the observations.

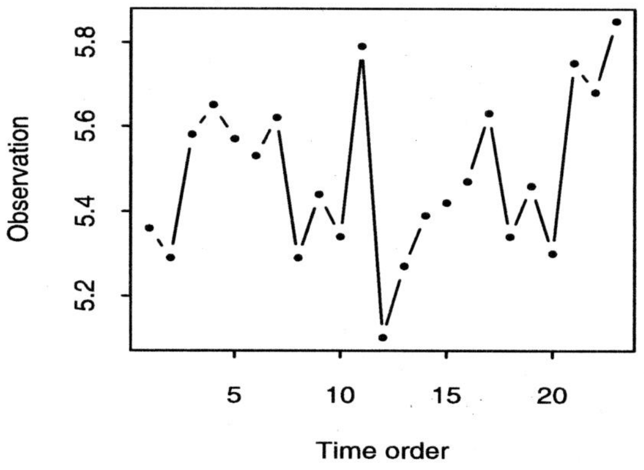

2.65 (a) The ordered data are

0.32 0.34 0.40 0.40 0.43 0.48 0.57

Since there are 7 observations, the median is the middle value. The median, maximum, minimum and range for the Tube 1 observations are:

Median = 0.40, maximum = 0.57, minimum = 0.32 and

range = maximum − minimum = 0.57 − 0.32 = 0.25.

(b) The ordered data are:

0.47 0.47 0.48 0.51 0.53 0.61 0.63

And the median, maximum, minimum and range for the Tube 2 observations are:

Median = 0.51, maximum = 0.63, minimum = 0.47 and

range = maximum − minimum = 0.63 − 0.47 = 0.16.

2.67 (a) Here, $11/4 = 2.25$ so Q_1 is the 3rd largest value. The quartiles for the velocity of light data are $Q_1 = 18.0$, $Q_2 = 27.0$, and $Q_3 = 30.0$

(b) The minimum, maximum, range and the interquartile range are

Minimum = 12, maximum = 48, range = maximum − minimum = 48 − 12 = 36 and

interquartile range $= Q_3 - Q_1 = 30.0 - 18.0 = 12.0$.

(c) The box-plot is

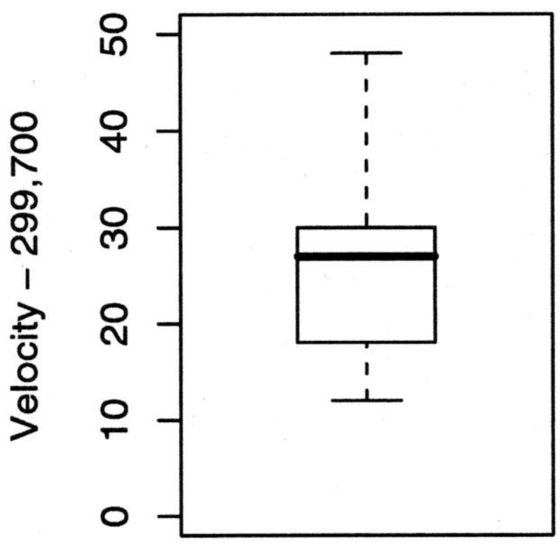

2.69 (a) The ordered data are: 12 14 19 20 21 28 29 30 55 63 63

The quartiles for the suspended solids data are $Q_1 = 19$, $Q_2 = 28$, and $Q_3 = 55$.

(b) The minimum, maximum, range and the interquartile range are

Minimum = 12, maximum = 63, range = 63 − 12 = 51 and

interquartile range = $Q_3 - Q_1 = 55 - 19 = 36$.

(c) The boxplot is

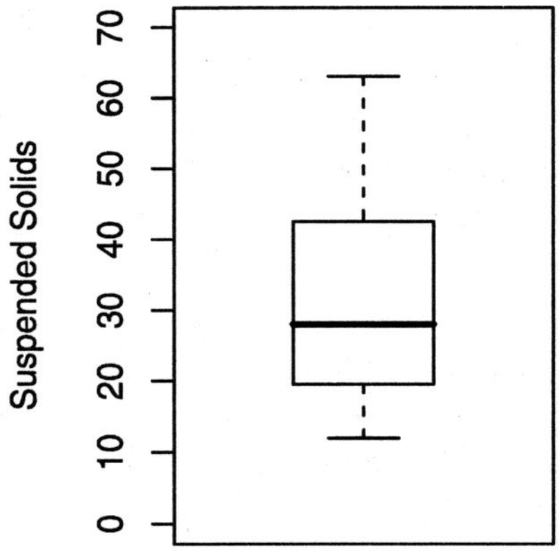

2.71 Boxplot of the aluminum data is given in Figure 2.1.

2.73 (a) $\overline{x} = 2.929/12 = .2441$

(b)

$$s^2 = \frac{\sum x_i^2 - (\sum x_i)^2/n}{n-1} = \frac{.715063 - (2.929)^2/12}{11} = .00001299$$

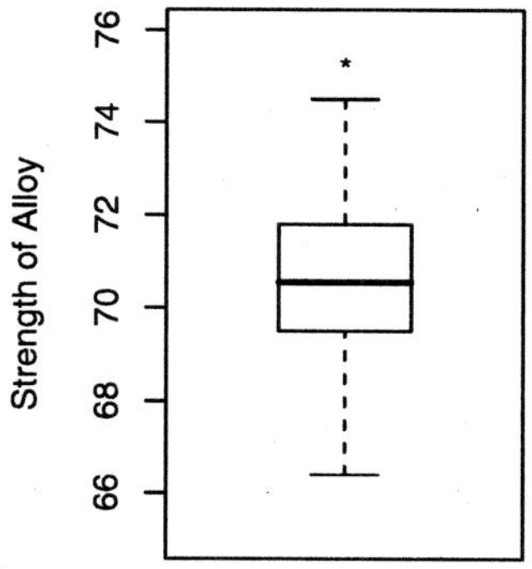

Figure 2.1: Boxplot for Exercise 2.71

so $s = \sqrt(.00001299) = .0036$

(c) The coefficient of variation is

$$\frac{.0036}{.2441} \cdot 100 = 1.47 \text{ percent.}$$

(d) For the large disk, the coefficient of variation is

$$\frac{.05}{.28} \cdot 100 = 17.9 \text{ percent.}$$

Thus, the values for the larger hard disk are relatively less consistent.

2.75 (a) The ordered observations are

389.1 390.8 392.4 400.1 425.9 429.1 448.4 461.6
479.1 480.8 482.9 497.2 505.8 516.5 517.5 547.5
550.9 563.7 567.7 572.2 572.5 575.6 595.5 602.0
606.7 611.9 618.9 626.9 634.9 644.0 657.6 679.3
698.6 718.5 738.0 743.3 752.6 760.6 794.8 817.2
833.9 889.0 895.8 904.7 986.4 1146.0 1156.0

The first quartile is the 12th observation, 497.2, the median is the 24th observation, 602.0, and the third quartile is the 36th observation, 743.3.

(b) Since $47(.90) = 42.3$, the 90th percentile is the 43rd observation, 895.8.

(c) The histogram is given in Figure 2.2

2.77 No. To be an outlier, the minimum or maximum must be quite separated from the next few closest observations.

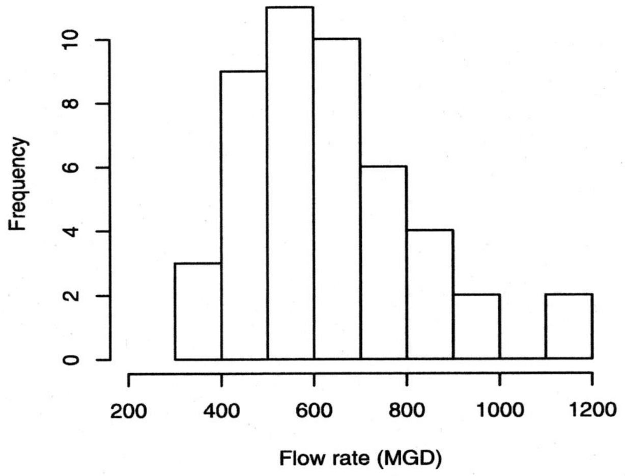

Figure 2.2: Histogram for Exercise 2.75(c).

2.79 (a) The dot diagram has a long right hand tail.

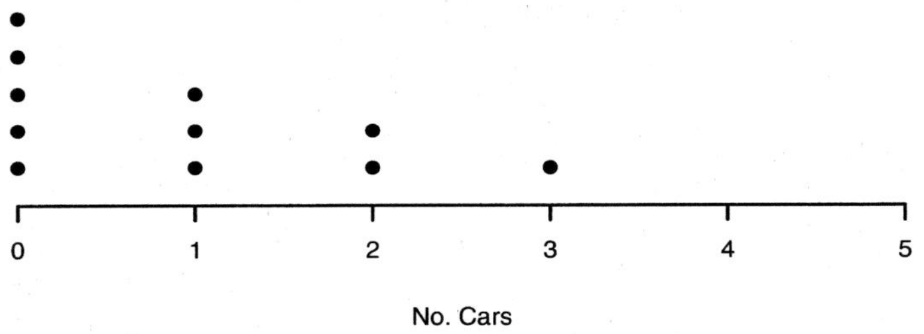

(b) The ordered data are

0 0 0 0 0 1 1 1 2 2 3

The median is the middle value, 1 car, and $\overline{x} = 10/11 = .909$ car which is smaller than the median.

Chapter 3

PROBABILITY

3.1 (a) A sketch of the 12 points of the sample space is as follows:

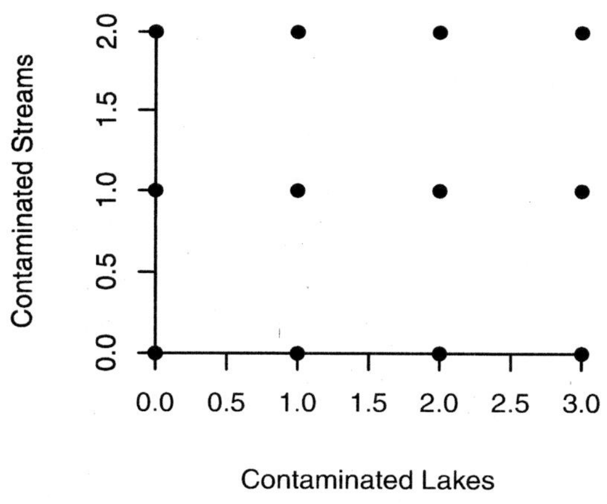

(b) $R=\{(0,0),(1,1),(2,2)\}$. $T=\{(0,0),(1,0),(2,0),(3,0)\}$. $U=\{(0,1),(0,2),(1,2)\}$.

3.3 (a) $R \cup U=\{(0,0),(1,1),(2,2),(0,1),(0,2),(1,2)\}$. $R \cup U$ is the event that the number of contaminated streams is greater than or equal to the number of contaminated lakes.

(b) $R \cap T=\{(0,0)\}$. $R \cap T$ is the event that none of the streams or lakes is contaminated.

(c) $\overline{T}=\{(0,1),(1,1),(2,1),(3,1),(0,2),(1,2),(2,2),(3,2)\}$. $\overline{T}$ is the event that at least one stream is contaminated.

3.5 $A=\{3,4\}$, $B=\{2,3\}$, $C=\{4,5\}$.

(a) $A \cup B=\{2,3,4\}$. Work is easy, average or difficult on this model.

(b) $A \cap B = \{3\}$. Work is average on this model.

(c) $\overline{B} = \{1,4,5\}$. Thus $A \cup \overline{B} = \{1,3,4,5\}$. Work is not easy on this model.

(d) $\overline{C} = \{1,2,3\}$. Work is very easy, easy or average on this model.

3.7 (a) A sketch of the 6 points of the sample space is given in Figure 3.1.

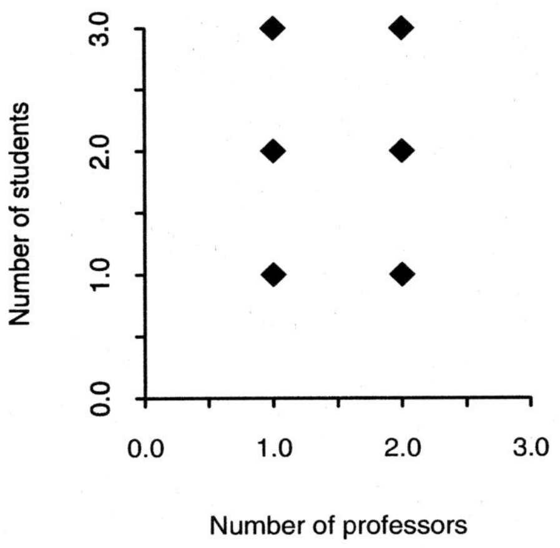

Figure 3.1: Sample space for Exercise 3.7.

(b) B is the event that 3 graduate students are present. C is the event that the same number of professors and graduate students are present. D is the event that the total number of graduate students and professors is 3.

(c) $C \cup D = \{(1,1),(1,2),(2,1),(2,2)\}$. $C \cup D$ is the event that at most 2 graduate students are present.

(d) B and D are mutually exclusive.

3.9 Region 1 is the event that the ore contains both uranium and copper. Region 2 is the event that the ore contains copper but not uranium. Region 3 is the event that the ore contains uranium but not copper. Region 4 is the event that the ore contains neither uranium nor copper.

3.11 (a) Region 5 represents the event that the windings are improper, but the shaft size is not too large and the electrical connections are satisfactory.

(b) Regions 4 and 6 together represent the event that the electrical connections are unsatisfactory, but the windings are proper.

(c) Regions 7 and 8 together represent the event that the windings are proper and the electrical connections are satisfactory.

(d) Regions 1, 2, 3, and 5 together represent the event that the windings are improper.

3.13 The following Venn diagram will be used in parts (a), (b), (c) and (d).

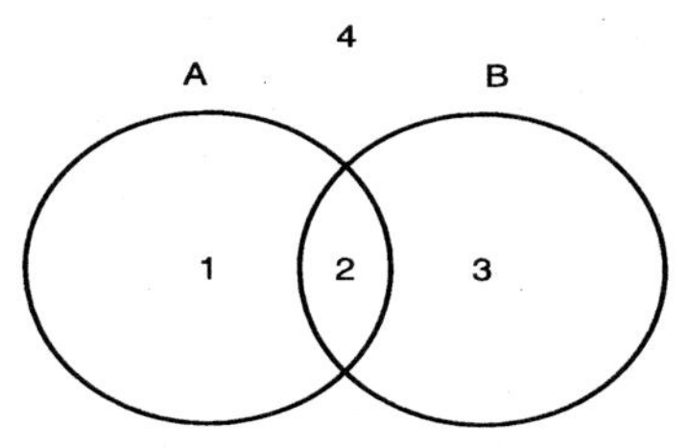

Fig. 1

(a) $A \cap B$ is region 2 in Fig. 1. $\overline{(A \cap B)}$ is the region composed of areas 1, 3, and 4. $\overline{A}$ is the region composed of areas 3 and 4. $\overline{B}$ is the region composed of areas 1 and 4. $\overline{A} \cup \overline{B}$ is the region composed of areas 1, 3, and 4. This corresponds to $\overline{(A \cap B)}$.

(b) $A \cap B$ is the region 2 in the figure. A is the region composed of areas 1 and 2. Since $A \cap B$ is entirely contained in A, $A \cup (A \cap B) = A$.

(c) $A \cap B$ is region 2. $A \cap \overline{B}$ is region 1. Thus, $(A \cap B) \cup (A \cap \overline{B})$ is the region composed of areas 1 and 2 which is A.

(d) From part (c), we have $(A \cap B) \cup (A \cap \overline{B}) = A$. Thus, we must show that $(A \cap B) \cup (A \cap \overline{B}) \cup (\overline{A} \cap B) = A \cup (\overline{A} \cap B) = A \cup B$. A is the region composed of areas 1 and 2 and $\overline{A} \cap B$ is region 3. Thus, $A \cup (\overline{A} \cap B)$ is the region composed of areas 1, 2, and 3.

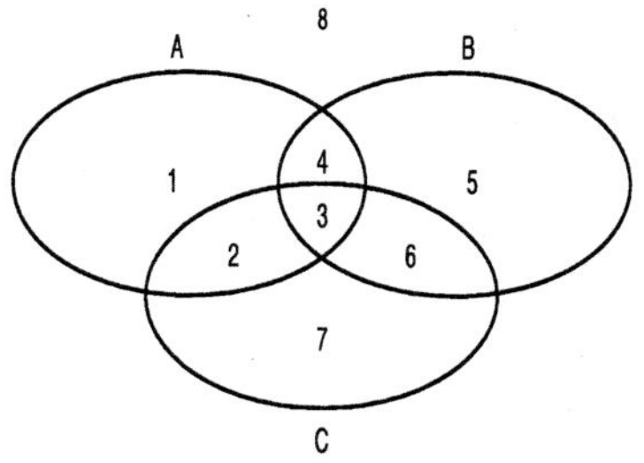

Fig. 2

(e) In Fig. 2, $A \cup B$ is the region composed of areas 1, 2, 3, 4, 5, and 6. $A \cup C$ is the region composed of areas 1, 2, 3, 4, 6, and 7, so $(A \cup B) \cap (A \cup C)$ is the region composed of areas 1, 2, 3, 4, and 6. $B \cap C$ is the region composed of areas 3, and 6, and A is the region composed of areas 1, 2, 3, and 4. Thus, $A \cup (B \cap C)$ is the region composed of areas 1, 2, 3, 4, and 6. Thus $A \cup (B \cap C) = (A \cup B) \cap (A \cup C)$.

3.15 The tree diagram is given in Figure 3.2, where S = Spain, U = Uruguay, P = Portugal and J = Japan.

3.17 There are $(6)(4)(3) = 72$ ways.

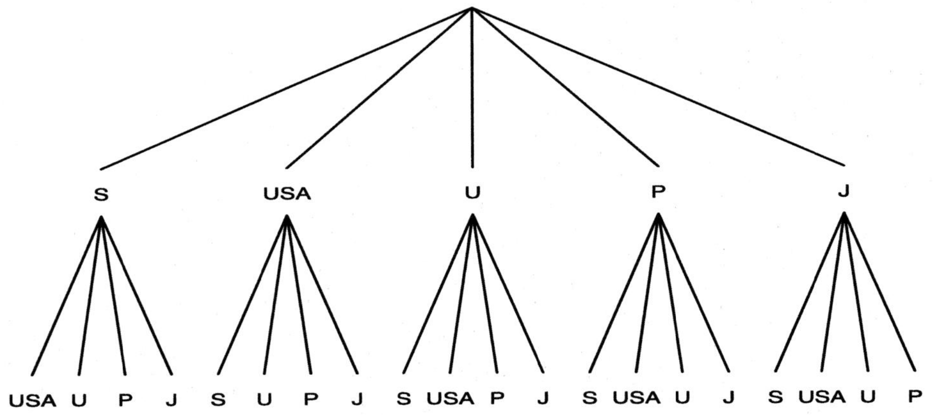

Figure 3.2: The tree diagram for Exercise 3.15.

3.19 (a) There men and women can be chosen in

$$_6C_2 = \binom{6}{2} = \frac{6!}{4!\,2!} = 15 \quad \text{and} \quad _4C_2 = \binom{4}{2} = \frac{4!}{2!\,2!} = 6$$

ways so there are $15 \times 6 = 90$ different project teams consisting of 2 men and 2 women.

(b) We are restricted from the choice of having the two women in question both selected giving a total of $6 - 1 = 5$ choices for two women. The number of project teams is reduced to $15 \times 5 = 75$.

3.21 $6! = 720$.

3.23 Since order does not matter, there are

$$_{15}C_2 = \binom{15}{2} = \frac{15!}{13!\,2!} = 105$$

ways.

3.25 There are $_{12}C_3 = 220$ ways to draw the three rechargeable batteries.
There are $_{11}C_3 = 165$ ways to draw none are defective.

(a) The number of ways to get the one that is defective is $220 - 165 = 55$.

(b) There are 165 ways not to get the one that is defective.

3.27 There are $_8C_2$ ways to choose the electric motors and $_5C_2$ ways to choose the switches. Thus, there are

$$_8C_2 \cdot {_5C_2} = 28 \cdot 10 = 280$$

ways to choose the motors and switches for the experiment.

3.29 The outcome space is

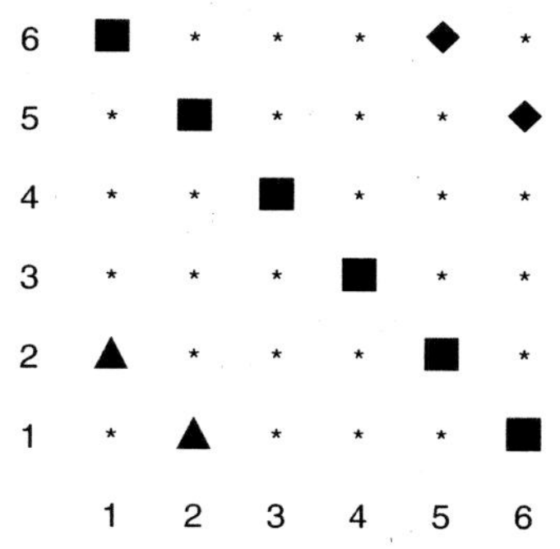

(a) The 6 outcomes summing to 7 are marked by squares. Thus, the probability is $6/36 = 1/6$.

(b) There are 2 outcomes summing to 11, which are marked by diamonds. Thus, the probability is $2/36 = 1/18$.

(c) These events are mutually exclusive. Thus, the probability is $6/36 + 2/36 = 2/9$.

(d) The 2 outcomes are marked by triangles. Thus, the probability is $2/36 = 1/18$.

(e) There are two such outcomes, (1,1) and (6,6). Thus, the probability is $2/36 = 1/18$.

(f) There are four such outcomes, (1,1), (1,2), (2,1) and (6,6). Thus, the probability is $4/36 = 1/9$.

3.31 There are $18 + 12 = 30$ cars. Thus, there are ${}_{30}C_4$ ways to choose the cars for inspection. There are ${}_{18}C_2$ ways to get the compacts and ${}_{12}C_2$ ways to get the intermediates. Thus, the probability is:

$$\frac{\binom{18}{2}\binom{12}{2}}{\binom{30}{4}} = \frac{10,098}{27,405} = .368.$$

3.33 The number of students enrolled in the statistics course or the operations research course is $92+63-40 = 115$. Thus, $160 - 115 = 45$ are not enrolled in either course.

3.35 (a) Yes. $P(A) + P(B) + P(C) + P(D) = 1$.

(b) No. $P(A) + P(B) + P(C) + P(D) = 1.02 > 1$.

(c) No. $P(C) = -.06 < 0$.

(d) No. $P(A) + P(B) + P(C) + P(D) = 15/16 < 1$.

(e) Yes. $P(A) + P(B) + P(C) + P(D) = 1$.

3.37 (a) There is 1 point where $i + j = 2$. There are 2 points where $i + j = 3$. There are 2 points where $i + j = 4$. There is 1 point where $i + j = 5$. Thus,

$$\frac{15/28}{2} + 2 \cdot \frac{15/28}{3} + 2 \cdot \frac{15/28}{4} + \frac{15/28}{5}$$
$$= \frac{15}{28} \cdot (1/2 + 2/3 + 1/2 + 1/5) = \frac{15}{28} \cdot \frac{28}{15} = 1.$$

Since each probability is between 0 and 1, the assignment is permissible.

(b) $P(B) = 15/28 \cdot (1/4 + 1/5) = 5/28 \cdot 9/20 = 135/560 = 27/112$.

$P(C) = 15/28 \cdot (1/2 + 1/4) = 15/28 \cdot 3/4 = 45/112$.

$P(D) = 15/28 \cdot (1/3 + 1/3) = 15/28 \cdot 2/3 = 5/14$.

(c) The probability that 1 graduate student will be supervising the lab is:

$$\frac{15/28}{2} + \frac{15/28}{3} = \frac{25}{56} = .446.$$

The probability that 2 graduate students will be supervising the lab is:

$$\frac{15/28}{3} + \frac{15/28}{4} = \frac{35}{112} = \frac{5}{16} = .3125.$$

The probability that 3 graduate students will be supervising the lab is:

$$\frac{15/28}{4} + \frac{15/28}{5} = \frac{27}{112} = .241.$$

3.39 (a) $(A \cap B) \cup (A \cap \overline{B}) = A$, and $A \cap B$ and $A \cap \overline{B}$ are disjoint. Thus

$$P((A \cap B) \cup (A \cap \overline{B})) = P(A \cap B) + P(A \cap \overline{B}) = P(A).$$

Since $P(A \cap \overline{B}) \geq 0$, we have proved that $P(A \cup B) \leq P(A)$.

(b) Combining (d) and (c) of Exercise 3.13 gives us

$$A \cup B = (A \cap B) \cup (A \cap \overline{B}) \cup (\overline{A} \cap B) = A \cup (\overline{A} \cap B).$$

But A and $\overline{A} \cap B$ are disjoint. Thus,

$$P(A \cup B) = P(A) + P(\overline{A} \cap B).$$

Since $P(\overline{A} \cap B) \geq 0$, we have proved that $P(A \cup B) \geq P(A)$.

3.41 (a) $P(\overline{A}) = 1 - P(A) = 1 - .26 = .74$.

(b) $P(A \cup B) = P(A) + P(B) = .26 + .45 = .71$, since A and B are mutually exclusive.

(c) $P(A \cap \overline{B}) = P(A) = .26$, since A and B are mutually exclusive.

(d) $P(\overline{A} \cap \overline{B}) = P(\overline{(A \cup B)}) = 1 - P(A \cup B) = 1 - .26 - .45 = .29$.

3.43 (a) This probability is given by $.22 + .21 = .43$.

(b) $.17 + .29 + .21 = .67$.

(c) $.03 + .21 = .24$.

(d) $.22 + .29 + .08 = .59$.

3.45 (a) 15/32 (b) 13/32 (c) 5/32 (d) 23/32 (e) 8/32 (f) 9/32.

3.47 (a) "At least one award" is the same as "design or efficiency award". Thus, the probability is $.16 + .24 - .11 = .29$.

(b) This the probability of "at least one award" minus the probability of both awards or $.29 - .11 = .18$.

3.49

$$P(A \cup B \cup C) = 1 - .11 = .89, \qquad P(A) = .24 + .06 + .04 + .16 = .5,$$
$$P(B) = .19 + .06 + .04 + .11 = .4, \qquad P(C) = .09 + .16 + .04 + .11 = .4,$$
$$P(A \cap B) = .06 + .04 = .1, \qquad P(A \cap C) = .16 + .04 = .2,$$
$$P(B \cap C) = .04 + .11 = .15, \qquad P(A \cap B \cap C) = .04.$$

Thus, the following equation must equal to .89:

$$.5 + .4 + .4 - .1 - .2 - .15 + .04 = .89.$$

This proves the formula.

3.51 (a) The odds for are $(4/7)/(3/7) = 4$ to 3.

(b) The odds against are $.95/.05 = 19$ to 1 against.

(c) The odds for are $.80/.20 = 4$ to 1.

3.53 (a) $p = 3/(3+2) = 3/5$.

(b) $30/(30+10) = 3/4 \leq p < 40/(10+40) = 4/5$.

3.55 $P(I \cap D) = 10/500$, $P(D) = 15/500$, $P(I \cap \overline{D}) = 20/500$, $P(\overline{D}) = 485/500$.

$$P(I|D) = \frac{P(I \cap D)}{P(D)} = \frac{10}{15} = \frac{2}{3},$$

$$P(I|\overline{D}) = \frac{P(I \cap \overline{D})}{P(\overline{D})} = \frac{20}{485} = \frac{4}{97}.$$

3.57 (a) The sample space is C. Thus the probability is given by:

$$\frac{N(A \cap C)}{N(C)} = \frac{8 + 15}{8 + 54 + 9 + 14} = \frac{62}{85} = .73.$$

(b) This is given by:

$$\frac{N(A \cap B)}{N(A)} = \frac{20 + 54}{20 + 54 + 8 + 2} = \frac{74}{84} = .881.$$

(c) This is given by:

$$\frac{N(\overline{C}\cap\overline{B})}{N(\overline{B})}=\frac{N(\overline{(C\cup B)})}{N(\overline{B})}=\frac{150-121}{105-99}=\frac{29}{51}=.569.$$

3.59 From the example on page 61, we know that $P(M_1)=0.20$, $P(P_1)=0.75$, $P(M_1\cap P_1)=0.16$, and $P(C_3)=0.40$

(a)

$$P(M_1|P_1)=\frac{0.16}{0.75}=.213 \quad \text{which is different from} \quad P(M_1)$$

(b) We first obtain $P(C_3\cap P_2)=0.01+0.06+0.02=0.09$ and

$$P(P_2)=0.02+0.01+0.01+0.02+0.07+0.06+0.01+0.03+0.02=0.25$$

Hence, $P(C_3|P_2)=0.09/0.25=.36$ which is different from $P(C_3)$.

(c) We first obtain $P(M_1\cap P_1\cap C_3)=0.07$ and $P(P_1\cap C_3)=.07+.10+.14=.31$ so that

$$P(M_1|P_1\cap C_3)=\frac{.07}{.31}=.2258 \quad \text{which is different from} \quad P(M_1)$$

3.61 $P(A|B)=P(A\cap B)/P(B)$ by definition. Thus, $P(A|B)=P(A)$ implies that $P(A\cap B)/P(B)=P(A)$, which implies $P(A\cap B)/P(A)=P(B)$, since both $P(A)$ and $P(B)$ are not zero. Thus $P(B|A)=P(B)$.

3.63 (a) $P(A|B)=P(A\cap B)/P(B)=.24/.40=.6=P(A)$.

(b) $P(A|\overline{B})=P(A\cap\overline{B})/P(\overline{B})=(P(A)-P(A\cap B))/(1-P(B))$
$=(.60-.24)/(1-.40)=.6=P(A)$.

(c) $P(B|A)=P(B\cap A)/P(A)=.24/.60=.4=P(B)$.

(d) $P(B|\overline{A})=P(B\cap\overline{A})/P(\overline{A})=(P(B)-P(B\cap A))/(1-P(A))$
$=(.40-.24)/(1-.60)=.4=P(B)$.

3.65 (a) The probability of drawing a Seattle-bound part on the first draw is 45/60. The probability of drawing a Seattle-bound part on the second draw given that a Seattle-bound part was drawn on the first draw is 44/59. Thus, the probability that both parts should have gone to Seattle is:

$$\frac{45}{60}\cdot\frac{44}{59}=.559.$$

(b) Using an approach similar to (a), the probability that both parts should have gone to Vancouver is:

$$\frac{15}{60}\cdot\frac{14}{59}=.059.$$

(c) The probability that one should have gone to Seattle and one to Vancouver is 1 minus the sum of the probability in parts (a) and (b) or .381.

3.67 A and B are independent if and only if $P(A)P(B) = P(A \cap B)$. Since (0.60)(0.45)= 0.27, they are independent.

3.69 (a) Each head has probability 1/2, and each toss is independent. Thus, the probability of 8 heads is $(1/2)^8 = 1/256$.

(b) $P(\text{three 3's and then a 4 or 5}) = (1/6)^3(1/3) = 1/648$.

(c) $P(\text{five questions answered correctly}) = (1/3)^5 = 1/243$.

3.71 Using the law of total probability, $P(\text{new worker meets quota}) = (.80)(.83) + (.20)(.35) = .734$.

3.73 $P(\text{car had bad tires}) = (.20)(.10)+(.20)(.12)+(.60)(.04)=.068$.

3.75 (a) $P(A) = (.4)(.3) + (.6)(.8) = .60$.

(b) $P(B|A) = P(B \cap A)/P(A) = (.4)(.3)/(.60) = .20$.

(c) $P(B|\overline{A}) = P(B \cap \overline{A})/P(\overline{A}) = (.4)(.7)/(.4) = .70$.

3.77 (a)

$$\begin{aligned} &P(\text{Tom} \mid \text{incomplete repair}) \\ &= \frac{(.6)(1/10)}{(.2)(1/20) + (.6)(1/10) + (.15)(1/10) + (.05)(1/20)} \\ &= \frac{.06}{.0875} = .686. \end{aligned}$$

(b)

$$P(\text{George} \mid \text{incomplete repair}) = \frac{(.15)(1/10)}{.0875} = .171.$$

(c)

$$P(\text{Peter} \mid \text{incomplete repair}) = \frac{(.05)(1/20)}{.0875} = .0286.$$

3.79 Let A be the event that the test indicates corrosion inside of the pipe and C be the event that corrosion is present. We are given $P(A|C) = .7$, $P(A|\overline{C}) = .2$, and $P(C) = .1$.

(a) By Bayes' theorem

$$\begin{aligned} P(C|A) &= \frac{P(A|C)P(C)}{P(A|C)P(C) + PA|\overline{C})P(\overline{C})} \\ &= \frac{.7 \times .1}{.7 \times .1 + .2 \times .9} = \frac{.07}{.07 + .18} = .28 \end{aligned}$$

(b)

$$P(C|\overline{A}) = \frac{P(\overline{A}|C)P(C)}{P(\overline{A}|C)P(C) + P(\overline{A}|\overline{C})P(\overline{C})}$$

$$= \frac{(1-.7)\times .1}{(1-.7)\times .1 + (1-.2)\times .9} = \frac{.03}{.03+.72} = .04$$

3.81 (a) Using the long run relative frequency approximation to probability, we estimate the probability

$$P\,[\,\text{Checked out}\,] = \frac{27}{300} = 0.09$$

(b) Using the data from last year, the long run relative frequency approximation to probability gives the estimate

$$P\,[\,\text{Get internship}\,] = \frac{28}{380} = 0.074$$

One factor might be the quality of permanent jobs that interns received last year or even how enthusiastic they were about the internship. Both would likely increase the number of applicants. Bad experiences may decrease the number of applicants.

3.83 (a) $\overline{A}$={(0,0),(0,1),(0,2),(0,3),(1,0),(1,1),(1,2),(2,0), (2,1),(3,0)}.
$\overline{A}$ is the event that the salesman will not visit all four of his customers.

(b) $A \cup B$={(4,0),(3,1),(2,2),(1,3),(0,4),(1,0),(2,0), (2,1),(3,0)}.
$A \cup B$ is the event that the salesman will visit all four customers or more on the first day than on the second day.

(c) $A \cap C$={(1,3),(0,4)}.
$A \cap C$ is the event that he will visit all four customers but at most one on the first day.

(d) $\overline{A} \cap B$={(1,0),(2,0),(2,1),(3,0)}.
$\overline{A} \cap B$ is the event that he will visit at most three of the customers and more on the first day than on the second day.

3.85 The tree diagram is given in Figure 3.3

3.87 There are ${}_7C_2 = 21$ ways to assign the chemical engineers.

3.89 (a) $P(A \cup B) = 0.35 + 0.40 - 0.20 = 0.55$.

(b) $P(\overline{A} \cap B) = P(B) - P(A \cap B) = 0.40 - 0.20 = 0.20$

(c) $P(A \cap \overline{B}) = P(A) - P(A \cap B) = 0.35 - 0.20 = 0.15$

(d) $P(\overline{A} \cup \overline{B}) = P(\overline{A \cap B}) = 1 - P(A \cap B) = 1 - .20 = 0.80$.

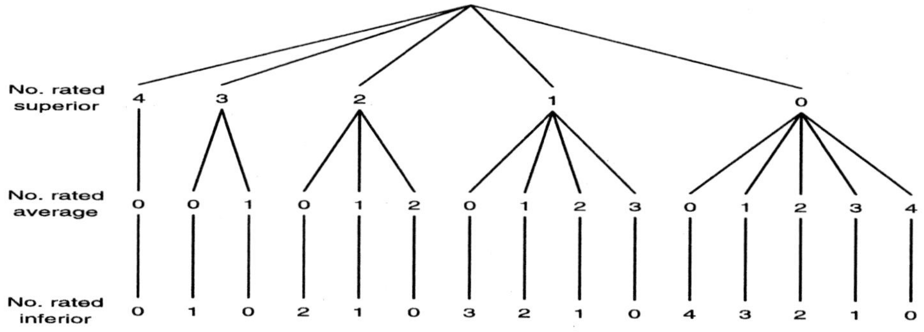

Figure 3.3: Tree diagram for Exercise 3.85.

(e) A and B are not independent, since

$$P(A)P(B) = (.35)(.4) = .14 \neq P(A \cap B).$$

3.91 The total number of that said the product was reliable or easy to use is $165 + 117 - 88 = 195$. If 33 said neither, then there must be $195 + 33 = 228$ in the survey , not 200.

3.93 (a) $P(A|B) = P(A \cap B)/P(B) = 0.09/0.45 = 0.2 = P(A)$.

(b) Since $P(A \cap \overline{B}) = P(A) - P(A \cap B) = 0.2 - 0.09 = 0.11$,

$$P(A|\overline{B}) = P(A \cap \overline{B})/P(\overline{B}) = \frac{0.11}{1 - 0.45} = .2 = P(A).$$

(c) $P(B|A) = P(B \cap A)/P(A) = 0.09/0.20 = 0.45 = P(B)$.

(d) Since $P(B \cap \overline{A}) = P(B) - P(B \cap A) = 0.45 - 0.09 = 0.36$,

$$P(B|\overline{A}) = P(B \cap \overline{A})/P(\overline{A}) = \frac{0.36}{1 - .2} = 0.45 = P(B).$$

3.95 Let us denote the chemicals Arsenic, Barium, and Mercury by the letters A, B, and M respectively, and indicate the concentrations by the subscripts 'H' for high and 'L' for low. For instance, a high concentration of Mercury will be denoted by M_H.

(a) Of the 58 landfills, the number with M_H is $1 + 4 + 5 + 10 = 20$.

Therefore, $P(M_H) = \frac{20}{58} = .344$

(b) The number of $M_L A_L B_H$ landfills is 8 so $P(M_L A_L B_H) = \frac{8}{58} = .138$

(c) There are three possibilities for landfills with two H's one L. The number of $A_H B_L M_H$ landfills is 5, the number of $A_L B_H M_H$ is 4, and the number of $A_H B_H M_L$ is 3, so the total is 12.

Therefore, $P(\text{two } H\text{'s and one } L\text{'s}) = \frac{12}{58} = .207$

(d) There are three possibilities for landfills with one H and two L's. The number of $A_H B_L M_L$ landfills is 9, the number of $A_L B_H M_L$ is 8, and the number of $A_L B_L M_H$ is 10, so the total is 27. Therefore, $P(\text{one } H \text{ and two } L\text{'s}) = \frac{27}{58} = .466$

3.97 Let events S.E. = static electricity, E = explosion, M = malfunction, O.F. = open flame, and P.E. = purposeful action. We need to find probabilities P(S.E.|E), P(M|E), P(O.F.|E), P(P.A.|E). Since

$$P(E) = (.30)(.25) + (.40)(.20) + (.15)(.40) + (.15)(.75) = .3275,$$

we have

$$P(S.E.|E) = (.30)(.25)/.3275 = .229, \qquad P(M|E) = (.40)(.20)/.3275 = .244,$$
$$P(O.F.|E) = (.15)(.40)/.3275 = .183, \qquad P(P.A.|E) = (.15)(.75)/.3275 = .344.$$

Thus, purposeful action is most likely.

3.99 We let A be the event route A is selected, B be the event route B is selected and C the event Amy arrives home at or before 6 p.m. We are given $P(A) = .4$ so $P(B) = 1 - .4 = .6$.

(a) By the law of total probability

$$P(C) = P(C|A)P(A) + P(C|B)P(B) = .8 \times .4 + \ .7 \times .6 = .74$$

(b) Using Bayes' rule

$$P(B|\overline{C}) = \frac{P(\overline{C}|B)P(B)}{P(\overline{C}|A)P(A) + P(\overline{C}|B)P(B)} = \frac{.3 \times .6}{.2 \times .4 + .3 \times .6} = .692$$

Chapter 4

PROBABILITY DISTRIBUTIONS

4.1 Let N be the total number of contaminated lakes and streams.

$$P(N=0) = 1/12 \qquad P(N=3) = 3/12 = 1/4$$
$$P(N=1) = 2/12 = 1/6 \qquad P(N=4) = 2/12 = 1/6$$
$$P(N=2) = 3/12 = 1/4 \qquad P(N=5) = 1/12.$$

The distribution of total number of contaminated lakes and streams can be tabulated as :

N	0	1	2	3	4	5
Prob	1/12	1/6	1/4	1/4	1/6	1/12

4.3 (a) Yes. $0 \leq f(i) \leq 1$, and $\sum_{i=1}^{4} f(i) \;=\; 1$.

(b) No. $\sum_{i=1}^{4} f(i) \;=\; 0.96 < 1$.

(c) No. $f(4) < 0$.

4.5 Using the identity

$$(x-1)\sum_{i=0}^{n} x^i \;=\; x^{n+1} - 1$$

or

$$\sum_{i=0}^{n} x^i \;=\; \frac{x^{n+1}-1}{x-1},$$

we have

$$\sum_{x=0}^{4} \frac{k}{2^x} \;=\; k\frac{(\frac{1}{2})^{4+1}-1}{\frac{1}{2}-1} \;=\; \frac{31k}{16}.$$

This must equal 1, so $k = 16/31$.

4.7

$$\begin{aligned}
b(x;n,p) &= \binom{n}{x} p^x(1-p)^{n-x} = \frac{n!}{x!\,(n-x)!}\, p^x(1-p)^{n-x} \\
&= \frac{n!}{(n-x)!\,x!}\,(1-p)^{n-x}\,p^x = \binom{n}{n-x}(1-p)^{n-x}\,p^x \\
&= b(n-x;n,1-p)
\end{aligned}$$

4.9 (a) Assumptions appear to hold. Success is a home with a TV tuned to mayor's speech. The probability of success is the proportion of homes around city having a TV tuned to the mayor's speech.

(b) The binomial assumptions do not hold because the probability of a serious violation for the second choice depends on which plant is selected first.

4.11 (a) Success is person has a cold. Colds are typically passed around in families so trials would not be independent. Therefore, the binomial distribution does not apply.

(b) Success means projector does not work properly. The binomial assumptions do not hold because the probability of a success for the second choice depends on which projector is selected first.

4.13 From Table 1:

(a) $B(7;19,.45) = .3169$.

(b) $b(7;19,.45) = .3169 - .1727 = .1442$.

(c) $B(8;10,.95) = .0861$.

(d) $b(8;10,.95) = .0861 - .0115 = .0746$.

(e) $\sum_{k=4}^{10} b(k;10,.35) = 1 - B(3;10,.35) = 1 - .5138 = .4862$.

(f) $B(4;9,.3) - B(1;9,.3) = .9012 - .1960 = .7052$.

4.15

$$b(2;4,.75) = \binom{4}{2}(.75)^2(.25)^{4-2} = .2109.$$

4.17 (a) $1 - B(11;\ 15,\ .7) = 1 - .7031 = .2969$.

(b) $B(6;\ 15,\ .7) = .0152$.

(c) $b(10;\ 15,\ .7) = B(10;\ 15,\ .7) - B(9;\ 15,\ .7) = .4845 - .2784 = .2061$.

4.19 (a) $P(18 \text{ are ripe}) = (.9)^{18} = .1501$.

(b) $1 - B(15; 18, .9) = 1 - .2662 = .7338.$

(c) $B(14; 18, .9) = .0982.$

4.21 (a) $B(2; 16, .05\ ) = .9571$

(b) $B(2; 16, .10\ ) = .7892$

(c) $B(2; 16, .15\ ) = .5614$

(d) $B(2; 16, .20\ ) = .3518$

4.23 (a) We use the hypergeometric distribution and the probabilities are

$$h(1; 2, 3, 4) = \frac{\binom{3}{1}\binom{1}{1}}{\binom{4}{2}} = \frac{3 \cdot 1}{2 \cdot 3} = .5 \quad \text{and} \quad h(2; 2, 3, 4) = \frac{\binom{3}{2}\binom{1}{0}}{\binom{4}{2}} = \frac{3 \cdot 1}{2 \cdot 3} = .5$$

(b)

$$F(x) = \begin{cases} 0, & x < 1 \\ .5, & 1 \le x < 2 \\ 1.0, & 2 \le x \end{cases}$$

(c) Both are shown in Figure 4.1. Note how the jumps of the cumultative distribution correspond to mass in the probability distribution.

4.25 Using the hypergeometric distribution,

(a)

$$h(0; 3, 6, 24) = \frac{\binom{6}{0}\binom{18}{3}}{\binom{24}{3}} = \frac{816}{2024} = .4032$$

(b)

$$h(1; 3, 6, 24) = \frac{\binom{6}{1}\binom{18}{2}}{\binom{24}{3}} = \frac{6 \cdot 153}{2024} = .4536$$

(c) The probability that at least 2 have defects is 1 minus the sum of the probabilities of none, and 1 having defects. $1 - (.4032 + .4536) = .1432$

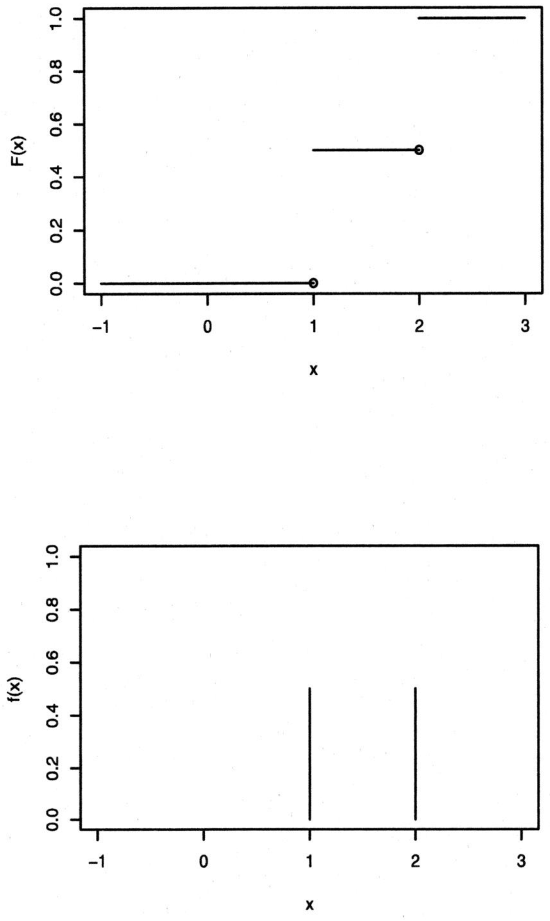

Figure 4.1: Cumulative Probability Distribution and Probability Histogram for Exercise 4.23.

4.27 (a)

$$P(\text{none in west}) = h(0; 3, 7, 16) = \frac{\binom{7}{0}\binom{9}{3}}{\binom{16}{3}} = \frac{1 \cdot 84}{560} = .15$$

(b)

$$P(\text{all in west}) = h(3; 3, 7, 16) = \frac{\binom{7}{3}\binom{9}{0}}{\binom{16}{3}} = \frac{35 \cdot 1}{560} = .0625$$

4.29 (a)

$$P(5 \text{ are union members}) = \frac{\binom{240}{5}\binom{60}{3}}{\binom{300}{8}} = .1470$$

(b)

$$P(5 \text{ are union members}) = \binom{8}{5}\left(\frac{240}{300}\right)^4\left(\frac{60}{300}\right)^3 = .1468$$

4.31 The cumulative binomial probabilities are

```
CDF;
BINOMIAL n = 27 p = .47 .

BINOMIAL WITH N =  27  P = 0.470000
   K  P( X LESS OR = K)
   2             0.0000
   3             0.0001
   4             0.0005
   5             0.0021
   6             0.0072
   7             0.0210
   8             0.0515
   9             0.1086
  10             0.1998
  11             0.3247
  12             0.4724
  13             0.6236
  14             0.7576
  15             0.8607
  16             0.9292
  17             0.9685
  18             0.9879
  19             0.9960
  20             0.9989
  21             0.9997
  22             1.0000
```

4.33 Using the computing formula:

$$\begin{aligned}\sigma^2 &= \mu_2' - \mu^2 \quad \text{with} \quad \mu = 1 \\ \mu_2' &= 0^2(.4) + 1^2(.3) + 2^2(.2) + 3^2(.1) = 2\end{aligned}$$

Thus,

$$\sigma^2 = 2 - 1^2 = 1.$$

4.35 Using the computing formula:

$$\sigma^2 = \mu_2' - \mu^2, \qquad \mu = 1.8$$

$$\mu_2' = 0^2(.17) + 1^2(.29) + 2^2(.27) + 3^2(.16) + 4^2(.07) + 5^2(.03) + 6^2(.01) = 5.04$$

Thus,

$$\sigma^2 = 5.04 - (1.8)^2 = 1.8.$$

The standard deviation is $\sigma = \sqrt{1.8} = 1.34$.

4.37 (a) The mean for the binomial distribution with $n = 4$ and $p = .7$ can be calculated from the following table:

i	0	1	2	3	4
$b(i; 4, .7)$	.0081	.0756	.2646	.4116	2401

Thus,

$$\begin{aligned} \mu &= 0(.0081) + 1(.0756) + 2(.2646) + 3(.4116) + 4(.2401) = 2.8 \\ \mu_2' &= 0^2(.0081) + 1^2(.0756) + 2^2(.2646) + 3^2(.4116) + 4^2(.2401) = 8.68 \\ \sigma^2 &= 8.68 - (2.8)^2 = .84. \end{aligned}$$

(b) $\mu = np = 4(.7) = 2.8$.
$\sigma^2 = np(1-p) = 4(.7\ )(.3) = .84$.

4.39 (a) The variance is given by:

$$\sigma^2 = (0-2.5)^2\frac{1}{32} + (1-2.5)^2\frac{5}{32} + (2-2.5)^2\frac{10}{32} + (3-2.5)^2\frac{10}{32}$$

$$+(4-2.5)^2\frac{5}{32} + (5-2.5)^2\frac{1}{32} = \frac{40}{32} = 1.25$$

(b) To use the computing formula we need:

$$\mu_2' = 0^2 \cdot \frac{1}{32} + 1^2 \cdot \frac{5}{32} + 2^2 \cdot \frac{10}{32} + 3^2 \cdot \frac{10}{32} + 4^2 \cdot \frac{5}{32} + 5^2 \cdot \frac{1}{32} = \frac{240}{32} = 7.5$$

Thus

$$\sigma^2 = 7.5 - (2.5)^2 = 1.25.$$

(c) The special formula for the binomial variance is:

$$\sigma^2 = np(1-p) = 5(.5)(.5) = 1.25$$

4.41 Since all of these random variables have binomial distributions,

(a) $\mu = np = 676(.5) = 338$ and $\sigma^2 = np(1-p) = 676(.5)(.5) = 169$ so $\sigma = 13$

(b)

$$\mu = np = 720 \cdot \frac{1}{6} = 120 \quad \textit{and} \quad \sigma^2 = np(1-p) = 720 \cdot \frac{1}{6}\frac{5}{6} = 100 \quad \text{so} \quad \sigma = 10$$

(c)

$$\mu = np = 600(.04) = 24 \quad \textit{and} \quad \sigma^2 = np(1-p) = 600(.04)(.96) = 23.04 \quad \text{so} \quad \sigma = 4.8$$

(d)

$$\mu = np = 800(.65) = 520 \quad \textit{and} \quad \sigma^2 = np(1-p) = 800(.65)(.35) = 182 \quad \text{so} \quad \sigma = 13.49$$

4.43 The mean of the hypergeometric distribution is

$$\begin{aligned}
\mu &= \sum_{x=0}^{n} x \frac{\binom{a}{x}\binom{N-a}{n-x}}{\binom{N}{n}} \\
&= \sum_{x=1}^{n} \frac{x\binom{a}{x}\binom{N-a}{n-x}}{\binom{N}{n}} \\
&= \frac{a}{\binom{N}{n}} \sum_{x=1}^{n} \binom{a-1}{x-1}\binom{N-a}{n-x}
\end{aligned}$$

Let $u = x - 1$. Then,

$$\mu = \frac{a}{\binom{N}{n}} \sum_{u=0}^{n-1} \binom{a-1}{u}\binom{N-a}{n-1-u}$$

Using the identity for the summation given in the problem, we have

$$\mu = \frac{a\binom{N-1}{n-1}}{\binom{N}{n}} = \frac{an}{N}$$

The result holds for all n such that $0 \leq n \leq N$, because

$$\binom{a}{x}\binom{N-a}{n-x} = 0 \quad \text{if} \quad x > a \quad \text{or} \quad (n-x) > (N-a)$$

4.45 (a) Since $\sigma = 0.002$ we have $0.006 = 3(0.0002)$, and $k = 3$.

$$P(\,|\,X - \mu\,| \leq 0.006) \;= 1 \;-\; P(\,|\,X - \mu\,| > 0.006) \geq 1 \;-\; \frac{1}{3^2} = \frac{8}{9}$$

(b) By the law of large numbers, this proportion will be close to the actual probability which is larger than the lower bound $8/9 = 0.889$.

4.47 We first find

$$\mu = 1{,}000{,}000 \cdot \frac{1}{2} = 500{,}000 \quad \text{and} \quad \sigma^2 = 1{,}000{,}000 \cdot \frac{1}{2} \cdot \frac{1}{2} = 250{,}000 \quad \text{so} \quad \sigma = 500.$$

If the proportion is between .495 and .505, the number of heads must be between 495,000 and 505,000. These bounds are both within 10 standard deviations of the mean. We can apply Chebyshev's theorem with $k = 10$. Thus, the probability is greater than $1 - 1/100 = .99$.

4.49 (a) By definition of variance,

$$\begin{aligned}
\sigma^2 &= \sum_{\text{all } x} (x-\mu)^2 f(x) \\
&= \sum_{\text{all } x} (x^2 - 2x\mu + \mu^2) f(x) \\
&= \sum_{\text{all } x} x^2 f(x) - \sum_{\text{all } x} 2x\mu f(x) + \sum_{\text{all } x} \mu^2 f(x) \\
&= \mu_2' - 2\mu \sum_{\text{all } x} x f(x) + \mu^2 \sum_{\text{all } x} f(x) \\
&= \mu_2' - 2\mu \cdot \mu + \mu^2 \cdot 1 \\
&= \mu_2' - \mu^2
\end{aligned}$$

(b) Similarly,

$$\begin{aligned}
\mu_3 &= \sum_{\text{all } x} (x-\mu)^3 f(x) \\
&= \sum_{\text{all } x} (x^3 - 3x^2\mu + 3x\mu^2 - \mu^3) f(x) \\
&= \sum_{\text{all } x} x^3 f(x) - \sum_{\text{all } x} 3x^2\mu f(x) + \sum_{\text{all } x} 3x\mu^2 f(x) - \sum_{\text{all } x} \mu^3 f(x) \\
&= \mu_3' - 3\mu \sum_{\text{all } x} x^2 f(x) + 3\mu^2 \sum_{\text{all } x} x f(x) - \mu^3 \sum_{\text{all } x} f(x) \\
&= \mu_3' - 3\mu \cdot \mu_2' + 3\mu^2 \cdot \mu - \mu^3 \cdot 1 \\
&= \mu_3' - 3\mu\mu_2' + 2\mu^3
\end{aligned}$$

4.51 For $\lambda = 3$, $f(0;\lambda) = e^{-3} = .0498$. Thus, using

$$f(x+1;\lambda) = \frac{\lambda}{x+1} f(x;\lambda)$$

$$f(1;\lambda) = \frac{\lambda}{0+1} f(0;\lambda) = 3e^{-3} = .1494.$$

$$f(2;\lambda) = \frac{3}{2} \cdot 3e^{-3} = .2240.$$

and so forth. The values are given in the following table:

x	0	1	2	3	4
$f(x;3)$	.0498	.1494	.2240	.2240	.1680

x	5	6	7	8	9
$f(x;3)$	.1008	.0504	.0216	.0081	.0027

The probability histogram is given in Figure 4.2.

4.53 (a) $F(9;12) = .242$

(b) $f(9;12) =$ F(9; 12) − F(8; 12) = .242 − .155 = .087

(c)

$$\sum_{k=3}^{12} f(k;7.5) = F(12;7.5) - F(2;7.5) = .9573 - .0203 = .937$$

where we have interpolated, by eye, between the entries for $\lambda = 7.4$ and $\lambda = 7.6$.

4.55 (a) $n = 80$, $p = .06$, $np = 4.8$. Thus,

$$f(4;8.4) = F(4;4.8) - F(3;4.8) = .476 - .294 = .182$$

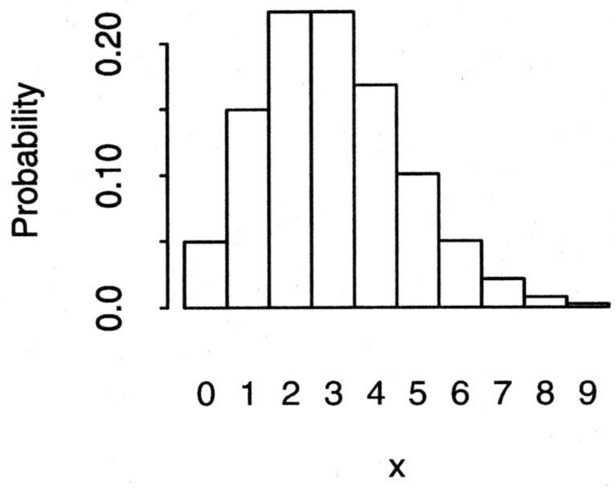

Figure 4.2: Probability Histogram for Exercise 4.51.

(b) $1 - F(2; 4.8) = 1 - .143 = .857$
(c)

$$\sum_{k=3}^{6} f(k; 4.8) = F(6; 4.8) - F(2; 4.8) = .791 - .143 = .648$$

4.57 $1 - F(12; 5.8) = 1 - .993 = .007$.

4.59 (a) $P(\text{at most 4 in a minute}) = F(4;\ 1.5\ ) = .981$.

(b) $P(\text{at least 3 in 2 minutes}) = 1 - F(2; 3) = 1 - .423 = .577$.

(c) $P(\text{at most 15 in 6 minutes}) = F(15; 9) = .978$.

4.61 P (fails after 1,200 times)

$$= \sum_{x=1201}^{\infty} (1-p)^{x-1} p = \frac{(1-p)^{1200}\, p}{1-(1-p)} = (1-p)^{1200}$$

where $p = .001$. Thus,

$$P(\text{fails after 1,200 times }) = (.999)^{1200} = .3010.$$

4.63 The required probability, given by the geometric distribution with $p = 0.10$, is

$$g(8; 0.1) = (0.9)^7 (0.1)^1 = 0.0478$$

4.65 We assume that the Poisson process with $\alpha = 0.01$ per hour applies.

(a) For a 4 hour time period, $\lambda = 0.01(4) = 0.04$.

$$f(1; 0.04) = \frac{(.04)^1}{1!} e^{-0.04} = 0.0384.$$

(b) We calculate $1 - f(0; 0.04) = 1 - e^{-0.04} = 0.0392$, or using Table 2,

$$1 - F(0; 0.04) = 1 - 0.961 = 0.039.$$

(c) For either of the two 4 hour time spans, the probability of exactly 1 customer is $f(1; 0.04) = 0.0384$. The two time intervals do not overlap so the counts are independent and we multiply the two probabilities

$$f(1; 0.04) \times f(1; 0.04) = (0.0384) \times (0.0384) = 0.0015$$

4.67 The Poisson process with $\alpha = 0.6$ applies.

(a) $\lambda = 0.6(2) = 1.2$.

$$f(2; 1.2) = \frac{(1.2)^2}{2!} e^{-1.2} = 0.2169.$$

(b) For the first hundred feet, the probability of exactly 1 flaw is

$$f(1; 0.6) = \frac{.06}{1} e^{-0.6} = .3293$$

The intervals do not overlap so the counts are independent and we multiply the two probabilities

$$f(1; 0.6) \times f(1; 0.6) = (.3293) \times (.3293) = .1084$$

4.69

$$\begin{aligned}
\mu &= \sum_{x=0}^{\infty} x f(x, \lambda) = \sum_{x=0}^{\infty} x \frac{(\lambda)^x}{x!} e^{-\lambda} = \lambda \sum_{x=1}^{\infty} \frac{(\lambda)^{x-1}}{(x-1)!} e^{-\lambda} \\
&= \lambda \sum_{x=0}^{\infty} \frac{(\lambda)^x}{x!} e^{-\lambda} \quad = \quad \lambda \\
\mu_2' &= \sum_{x=0}^{\infty} x^2 f(x, \lambda) = \sum_{x=0}^{\infty} x^2 \frac{(\lambda)^x}{x!} e^{-\lambda} = \sum_{x=1}^{\infty} \frac{x(\lambda)^x}{(x-1)!} e^{-\lambda} \\
&= \lambda \sum_{x=0}^{\infty} (x+1) \frac{(\lambda)^x}{x!} e^{-\lambda} \quad = \quad \lambda(\lambda + 1)
\end{aligned}$$

Thus,

$$\sigma^2 = \mu_2' - \mu^2 = \lambda(\lambda + 1) - \lambda^2 = \lambda$$

4.71 The cumulative Poisson probabilities are

(a)

```
Poisson with mu = 2.73000

     x     P( X <= x )
  2.00          0.4863
  3.00          0.7075
```

(b)

```
Poisson with mu = 4.33000

     x     P( X <= x )
  2.00          0.1936
  3.00          0.3718
```

4.73 To find this probability, we use the multinomial distribution with $n = 6$, $x_1 = 2$, $x_2 = 3$, $x_3 = 1$, $p_1 = 1/4$, $p_2 = 1/2$ and $p_3 = 1/4$, Thus the probability is given by

$$\frac{6!}{2!\,3!\,1!}\left(\frac{1}{4}\right)^2 \left(\frac{1}{2}\right)^3 \left(\frac{1}{4}\right)^1 = .117$$

4.75 (a) Let $\sum_{i=1}^{k} a_i = N$ and $\sum_{i=1}^{k} x_i = n$. Then the formula for the probability of getting x_i objects of the i-th kind when there are a_i objects of the i-th kind, for $i = 1, 2, ..., k$, is

$$\frac{\binom{a_1}{x_1}\binom{a_2}{x_2}\cdots\binom{a_k}{x_k}}{\binom{N}{n}} = \frac{\prod_{i=1}^{k}\binom{a_i}{x_i}}{\binom{N}{n}}$$

(b) Here $a_1 = 10$, $a_2 = 7$, $a_3 = 3$, with $n = 6$, $x_1 = 3$, $x_2 = 2$, and $x_3 = 1$. Thus the probability is

$$\frac{\binom{10}{3}\binom{7}{2}\binom{3}{1}}{\binom{20}{6}} = \frac{120 \cdot 21 \cdot 3}{38,760} = .1950$$

4.77 (a) The following table distributes the random numbers:

Number of sales	Probability	Cumulative probability	Random numbers
0	.14	.14	00–13
1	.28	.42	14–41
2	.27	.69	42–68
3	.18	.87	69–86
4	.09	.96	87–95
5	.04	1.00	96–99

Using columns 1 and 2 of Table 7, and starting in row 1, gives:

Day	1	2	3	4	5	6	7	8	9	10	11	12	13
Random no.	13	04	65	79	76	51	29	14	36	65	99	48	82
No. of sales	0	0	2	3	3	2	1	1	1	2	5	2	3

Day	14	15	16	17	18	19	20	21	22	23	24	25
Random no.	87	13	22	35	58	11	14	33	11	74	84	83
No. of sales	4	0	1	1	2	0	1	1	0	3	3	3

(b) The event of total sales greater than 9 on three consecutive days in a 25 day period is somewhat complicated. There are 23 sets of three consecutive days in this period and some dependence as all overlap two other cases. In our first trial, the total is 10 for days 11, 12 and 13 so the event of interest has occurred. We generated 99 sets of 25 days using the MINITAB command **Calc** > **Random Data** > **Discrete ....** A 1 denotes at least one total greater than 9 during one 25 day period. The first entry is From part (a).

```
1 0 1 0 1 1 1 0 1 1 1 1 1 0 1 1 1 1 1
1 1 1 0 1 1 1 1 0 1 0 1 0 1 1 1 0 1 1
1 1 1 0 0 1 0 1 0 1 1 1 1 1 1 1 1 1 1
0 0 1 1 1 0 0 1 0 1 1 0 0 1 1 0 1 0 1
1 0 1 1 1 0 1 1 1 1 0 0 1 1 1 1 1 0 1
1 1 0 0 1
```

Of the total of 100 trials of 25 days each, the event that some total for three consecutive days exceed 9, occurred 71 times. Therefore we estimate the probability as

$$P(\text{ Total greater than 9 for some three consecutive day period }) = \frac{71}{100} = 0.710$$

4.79 (a) $P(\text{ 2 or more defects }) = f(2) + f(3) = .03 + .01 = .04.$

(b) 0 is more likely since its probability $f(0) = .89$ is much larger than that of its complement $1 - .89 = .11$.

4.81 (a) $\mu = 0 \times .07 + 1 \times .15 + 2 \times .45 + 3 \times .25 + 4 \times .08 = 2.12.$

(b) We first calculate

$$0^2 \times .07 + 1^2 \times .15 + 2^2 \times .45 + 3^2 \times .25 + 4^2 \times .08 = 5.48$$

so variance $= 5.48 - (2.12)^2 = .9856$

(c) standard deviation $= \sqrt{.9856} = .9928$ rooms

4.83 (a) Yes. $0 \leq f(i) \leq 1$, and $\sum_{i=0}^{4} f(i) = 1$.

(b) Yes. $0 \leq f(i) \leq 1$, and $\sum_{i=-1}^{1} f(i) = 1$.

(c) No. $\sum_{i=0}^{3} f(i) \;=\; 1.5 > 1$

4.85 (a)

$$b(3;8,.2) = \binom{8}{3} (.2)^3\,(.8)^5 = \frac{8!}{3!\;5!}\,(.2)^3\,(.8)^5 = .1468$$

(b) $B(3;8,.2) - B(2;8,.2) = .9437 - .7969 = .1468$

4.87 Using the hypergeometric distribution,

(a)

$$h(0;2,4,16) = \frac{\binom{4}{0}\binom{12}{2}}{\binom{16}{2}} = \frac{1\cdot 66}{120} = .55$$

(b)

$$h(1;2,4,16) = \frac{\binom{4}{1}\binom{12}{1}}{\binom{15}{2}} = \frac{4\cdot 12}{120} = .40$$

(c)

$$h(2;2,4,16) = \frac{\binom{4}{2}\binom{12}{0}}{\binom{16}{2}} = \frac{6\cdot 1}{120} = .05$$

4.89 (a) The variance is given by:

$$\begin{aligned}\sigma^2 &= (0-1.2)^2(.216) + (1-1.2)^2(.432) + (2-1.2)^2(.288) + (3-1.2)^2(.064)\\ &= .72\end{aligned}$$

(b) Using the special formula for the binomial variance

$$\sigma^2 = np(1-p) = 3(.4)(.6) = .72$$

4.91 Here $n = 100$, $p = 0.02$ so $np = 2$. Using $\lambda = 2$, the approximate probability is $f(1;2)) = 2e^{-2}/1! = 0.2707$. Alternatively

$$f(1;2) = F(1;2) - F(1;2) = .406 - .135 = .271.$$

4.93 Since $(202 - 142)/12 = (142 - 82)/12 = 5$, we can apply Chebyshev's theorem with $k = 5$. Let X be the number of orders filled. Then,

$$P(X \leq 82 \text{ or } X \geq 202) = P(|X - 142| \geq 5 \cdot 12) \leq \frac{1}{25}$$

Thus,

$$P(82 < X < 202) > \frac{24}{25} = .96$$

4.95 $\lambda = 0.6$ for three weeks. The probability is

$$f(0;6) = (.6)^0 e^{-.6}/0! = .5488.$$

4.97 (a) The random numbers are distributed in the following table:

Number of spills	Probability	Cumulative probability	Random numbers
0	.2466	.2466	0000–2465
1	.3452	.5918	2466–5917
2	.2417	.8335	5918–8334
3	.1128	.9463	8335–9462
4	.0395	.9858	9463–9857
5	.0111	.9969	9858–9968
6	.0026	.9995	9969–9994
6	.0005	1.0000	9995–9999

(b) Using columns 9-12 of the third page of Table 7 and starting from row 101, gives:

Day	1	2	3	4	5
Random no.	8353	6862	0717	2171	3763
No. of spills	3	2	0	0	1

Day	6	7	8	9	10
Random no.	1230	6120	3443	9014	4124
No. of spills	0	2	1	3	1

Day	11	12	13	14	15
Random no.	7299	0127	5056	0314	9869
No. of spills	2	0	1	0	5

Day	16	17	18	19	20
Random no.	6251	4972	1354	3695	8898
No. of spills	2	1	0	1	3

Day	21	22	23	24	25
Random no.	1516	8319	3687	6065	3123
No. of spills	0	2	1	2	1

Day	26	27	28	29	30
Random no.	4802	8030	6960	1127	7749
No. of spills	1	2	2	0	2

Chapter 5

PROBABILITY DENSITIES

5.1

$$f(x) = \begin{cases} 2e^{-2x} & \text{for } x > 0 \\ 0 & \text{elsewhere} \end{cases}$$

Since $2e^{-2x}$ is positive on $x > 0$, $f(x)$ is always ≥ 0.

$$\int_{-\infty}^{\infty} f(x)dx = -e^{-2x}\Big|_0^{\infty} = 1$$

Thus, $f(x)$ is a density.

5.3 The distribution function is given by

$$F(x) = \int_{-\infty}^{x} f(s)ds = x^3$$

(a) $P(X > .8) = 1 - F(.8) = .488$

(b) $P(.2 < X < .4) = F(.4) - F(.2) = .056$

5.5

$$\begin{aligned} F(x) &= \int_{-\infty}^{x} f(s)ds = \begin{cases} 0 & x < 0 \\ x^2/2 & 0 \leq x \leq 1 \\ 1/2 + [2s - s^2/2]\big|_1^x & 1 < x \leq 2 \\ 1 & x > 2 \end{cases} \\ &= \begin{cases} 0 & x < 0 \\ x^2/2 & 0 \leq x \leq 1 \\ 2x - x^2/2 - 1 & 1 < x \leq 2 \\ 1 & x > 2 \end{cases} \end{aligned}$$

(a) $P(X > 1.8) = 1 - F(1.8) = 1 - [2(1.8) - (1.8)^2/2 - 1] = 1 - .98 = .02$

(b) $P(.4 < X < 1.6) = F(1.6) - F(.4) = 2(1.6) - (1.6)^2/2 - 1 - (.4)^2/2 = .84$

5.7 Let X have distribution $F(x)$. Then,

(a) $P(X < 3) = F(3) = 1 - 1/9 = 8/9 = 0.8889$

(b) $P(4 \leq X \leq 5) = F(5) - F(4) = 1/16 - 1/25 = 0.0225$

5.9 (a)

$$P(0 \leq \text{phase error} \leq \pi/4) = \int_0^{\pi/4} \cos x dx = \sin x|_0^{\pi/4} = \sin(\pi/4) = \sqrt{2}/2 = 0.707$$

(b) $P(\text{phase error} > \pi/3) = \int_{\pi/3}^{\pi/2} \cos x dx = \sin(\pi/2) - \sin(\pi/3) = 1 - \sqrt{3}/2 = .1339$

5.11 Integrating the density function by parts shows that the distribution function is given by

$$F(x) = 1 - \frac{1}{3}xe^{-x/3} - e^{-x/3}$$

Thus, the probability that the power supply will be inadequate on any given day is

$$P(\text{ power supply inadequate }) = P(\text{consumption} \geq 12 \text{ million kwh's})$$

$$= 1 - F(12) = 4e^{-4} + e^{-4} = 5e^{-4} = .0916$$

5.13 The density is

$$f(x) = \begin{cases} 3x^2 & 0 < x < 1 \\ 0 & \text{elsewhere} \end{cases}$$

Thus,

$$\mu = \int_0^1 3x^3 dx = 3x^4/4\big|_0^1 = \frac{3}{4} = 0.75$$

$$\mu_2' = \int_0^1 3x^4 dx = 3x^5/5\big|_0^1 = \frac{3}{5} = 0.6$$

and the variance is

$$\sigma^2 = \mu_2' - \mu^2 = 0.6 - (0.75)^2 = 0.0375$$

5.15 The density is:

$$f(x) = \begin{cases} 8x^{-3} & x > 2 \\ 0 & x \leq 2 \end{cases}$$

Thus,

$$\mu = \int_2^\infty x(8x^{-3})dx = -8x^{-1}\big|_2^\infty = 4$$

and

$$\mu_2' = \int_2^\infty x^2(8x^{-3})dx = 8 \ln x\,\big|_2^\infty = \infty$$

Thus, σ^2 does not exist.

5.17 The density is:

$$f(x) = \begin{cases} (1/4.5)e^{-x/4.5} & x > 0 \\ 0 & x \leq 0 \end{cases}$$

Thus,

$$\mu = \frac{1}{4.5}\int_0^\infty xe^{-x/4.5}dx$$

Integrating by parts gives:

$$\begin{aligned} \mu &= -xe^{-x/4.5}\Big|_0^\infty + \int_0^\infty e^{-x/4.5}dx \\ &= 0 - 4.5e^{-x/4.5}\Big|_0^\infty \\ &= 4.5 \text{ (years)} \end{aligned}$$

5.19 (a) $P(\text{less than } 1.75) = F(1.75) = .9599$

(b) $P(\text{less than } -1.25) = F(-1.25) = .1056$

(c) $P(\text{greater than } 2.06) = 1 - F(2.06) = 1 - .9803 = .0197$

(d) $P(\text{greater than } -1.82) = F(1.82) = .9656$

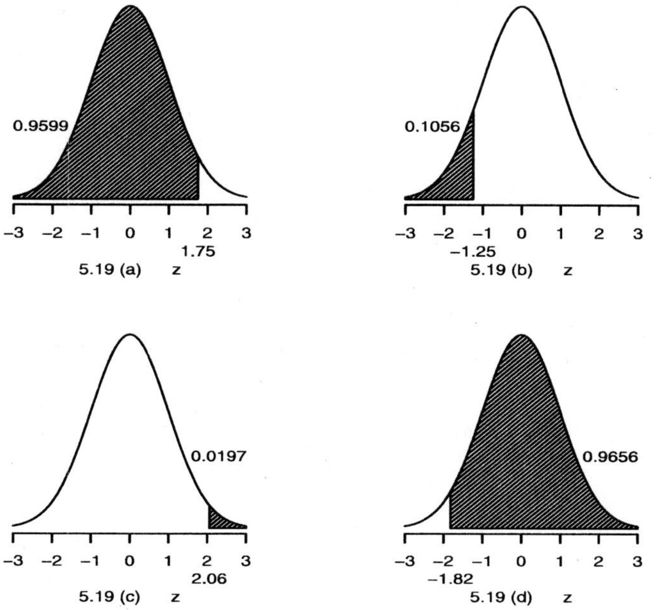

5.21 (a) $P(Z \leq z) = F(z) = .9911$. Thus $z = 2.37$

(b) $P(Z > z) = .1093$. That is, $P(Z \leq z) = 1 - .1093$ or $F(z) = .8907$. Thus, $z = 1.23$

(c) $P(Z > z) = .6443$. That is, $F(z) = 1 - .6443 = .3557$. Using Table 3, $z = -.37$

(d) $P(Z < z) = .0217$ so z is negative. From Table 3, $z = -2.02$.

(e) $P(-z \le Z \le z) = .9298$. That is, $F(z) - F(-z) = .9298$, which implies that $F(z) - (1 - F(z)) = .9298$ or $F(z) = (1 + .9298)/2 = .9649$. By Table 3, $z = 1.81$.

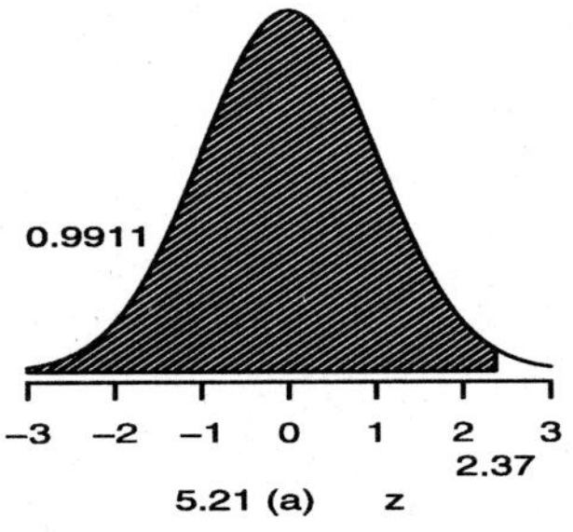

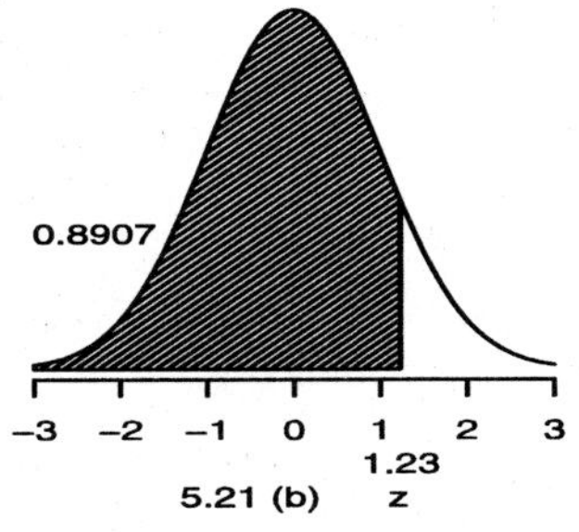

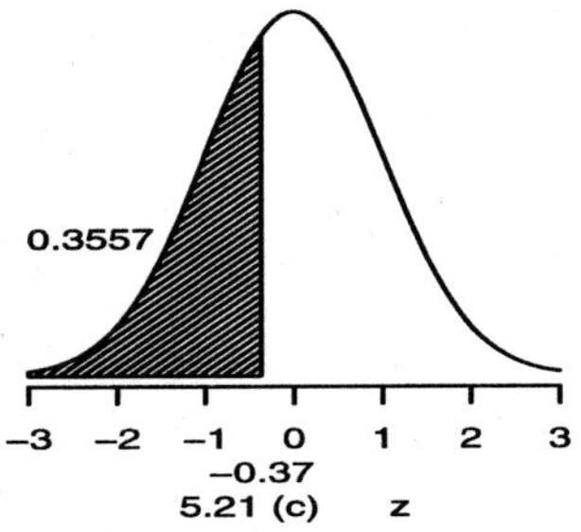

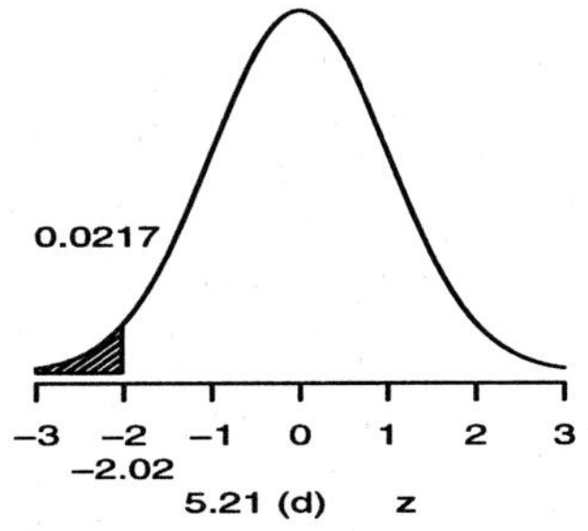

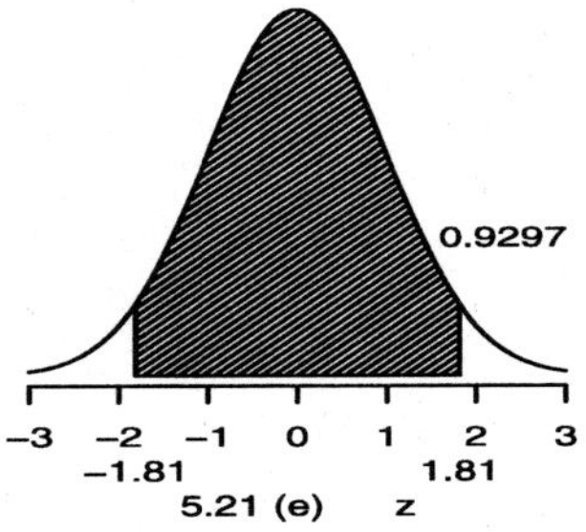

5.23 (a) $P(Z > z_{.005}) = .005$. Thus, $F(z_{.005}) = .995$ and $z = 2.575$ by linear interpolation in the Table 3.

(b) $P(Z > z_{.025}) = .025$. Thus, $F(z_{.025}) = .975$ and $z = 1.96$

5.25

$$P[X > 39.2] = .20 \quad \text{so} \quad P[\frac{X - 30}{\sigma} > \frac{9.2}{\sigma}] = .20$$

That is, $1 - F(9.2/\sigma) = .20$, and $F(9.2/\sigma) = .80$. But $F(.842) = .80$. Thus $9.2/\sigma = .842$, so $\sigma = 10.93$.

5.27 (a) We need to find $P(X > 11.5)$, where X is normally distributed with $\mu = 12.9$ and $\sigma = 2$.

$$P(X \ge 11.5) \;=\; 1 - F((11.5 - 12.9)/2) \;=\; F((12.9 - 11.5)/2)$$

$$= F(.7) = .7580$$

(b)

$$\begin{aligned} P(11 \le X \le 14.8) &= F((14.8 - 12.9)/2) - F((11 - 12.9)/2) \\ &= F(.95) - F(-.95) = .8289 - .1711 = .6578 \end{aligned}$$

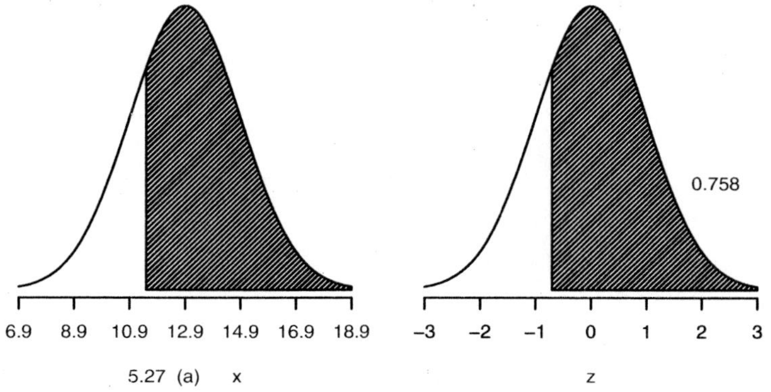

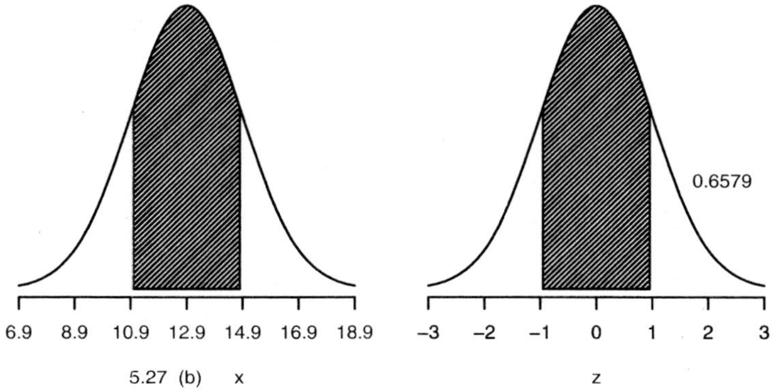

5.29 Let X be a random variable representing the processing time which is normally distributed with $\mu = 16.28$ and $\sigma = .12$.

(a) $P(10.00 \le X \le 10.50) = F((10.50 - 10.28)/.12) - F((10 - 10.28)/.12)$
$= F(1.833) - F(-2.333) = .9666 - .0098 = .9568$
(These values are determined by interpolation)

(b) $P(X \ge 10.20) = 1 - F((10.20 - 10.28)/.12) = 1 - F(-.667)$
$= F(.667) = .7476$

(c) $P(X \le 10.35) = F((10.35 - 10.28)/.12) = F(.5833) = .7201$

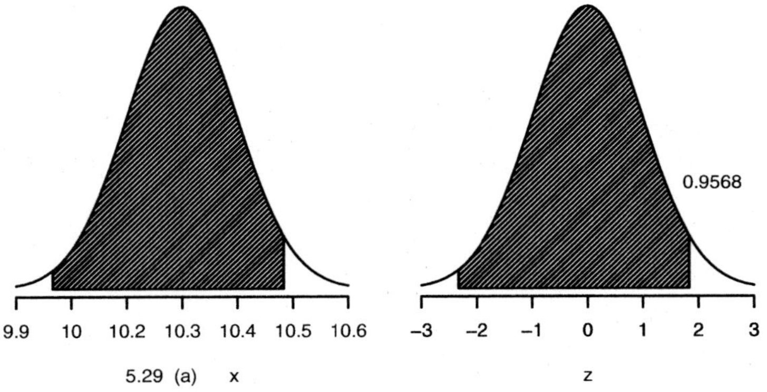

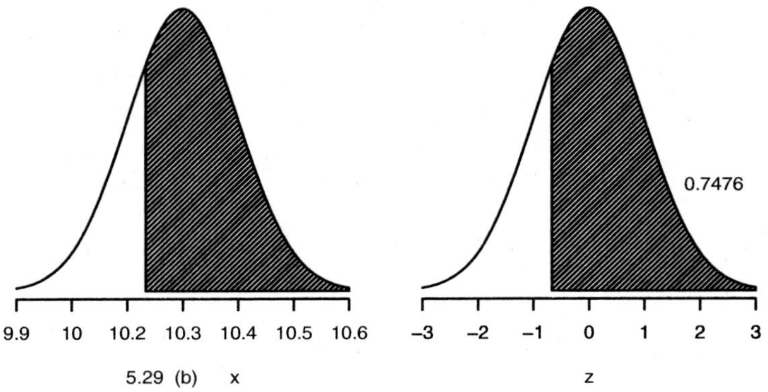

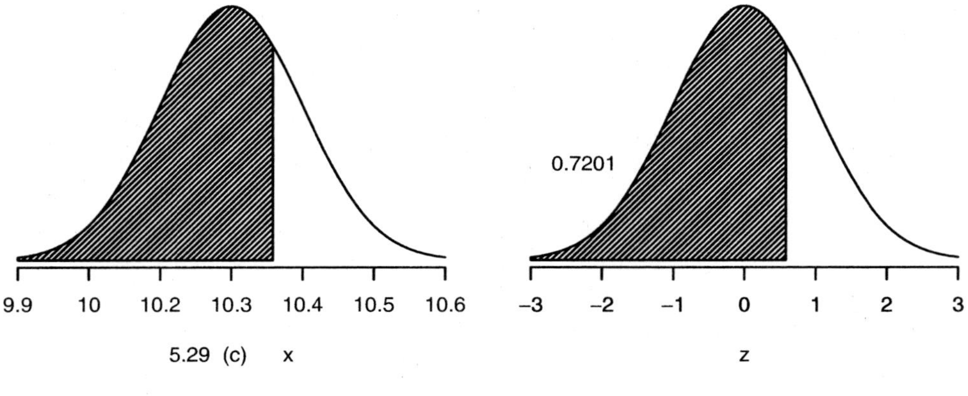

5.31 $P(.295 \leq X \leq .305) = F((.305 - .302)/.003) - F((.295 - .302)/.003)$

$= F(1) - F(-2.333) = .8413 - .0098 = .8315$

Thus, 83.15 percent will meet specifications.

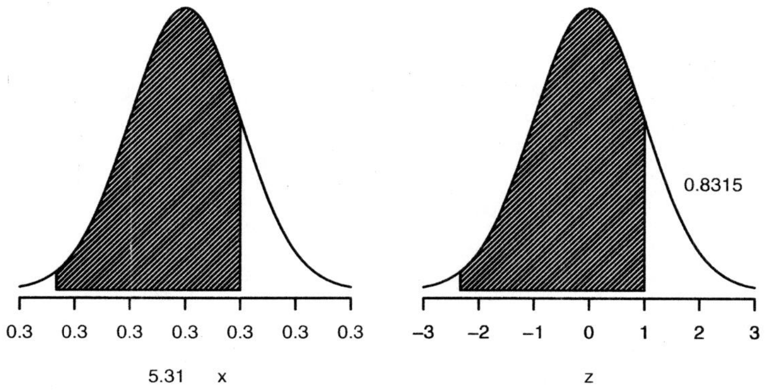

Problem 5.31

5.33 We need to find μ such that $F((3-\mu)/.01) = .95$. Thus, from Table 3, $(3-\mu)/.01 = 1.645$ or $\mu = 2.98355$.

5.35 If $n = 40$ and $p = .40$ then $\mu = 40(.40) = 16$ and $\sigma^2 = 40(.4)(.6) = 9.6$ or $\sigma = 3.0984$.

(a) $P(22) = F((22.5-16)/3.0984) - F((21.5-16)/3.0984)$
$= F(2.098) - F(1.775) = .9820 - .9621 = .0199$

(b) $P(\text{less than } 8) = F((7.5-16)/3.0984) = F(-2.743) = .0030$

5.37 In this case, $n = 200$, $p = .25$, $\mu = np = 50$, $\sigma^2 = np(1-p) = 37.5$, $\sigma = 6.1237$. Thus,

$$\begin{aligned} P(\text{fewer than 45 fail}) &= F((44.5-50)/6.1237) \\ &= F(-.90) \;=\; .1841 \end{aligned}$$

5.39 Again, we will use the normal approximation to the binomial distribution. Here, $n = 40$, $p = .62$, $\mu = np = 24.8$, $\sigma^2 = np(1-p) = 9.424$, $\sigma = 3.0699$. Thus,

$$F((20.5-24.8)/3.0699) = F(-1.40) \;=\; .0808$$

5.41 Let $f(x)$ be the standard normal density. Then $F(-z) = \int_{-\infty}^{-z} f(x)dx$. Using the change of variable, $s = -x$, and the fact that $f(x) = f(-x)$, we have

$$F(-z) = -\int_{\infty}^{z} f(-s)ds = \int_{z}^{\infty} f(s)ds = 1 - \int_{-\infty}^{z} f(s)ds = 1 - F(z)$$

5.43 We need to find

$$\begin{aligned} &\frac{1}{\sqrt{2\pi\sigma^2}} \int_{-\infty}^{\infty} (x-\mu)^2 \exp\left[-\frac{(x-\mu)^2}{2\sigma^2}\right] dx \\ &= \quad \frac{2}{\sqrt{2\pi\sigma^2}} \int_{\mu}^{\infty} (x-\mu)^2 \exp\left[-\frac{(x-\mu)^2}{2\sigma^2}\right] dx \end{aligned}$$

since the integrand is an even function. Using the change of variable $s = (x - \mu)/\sigma$, the variance is equal to

$$\frac{2\sigma^2}{\sqrt{2\pi}} \int_0^\infty s^2 \exp[-s^2/2] ds$$

Integrating by parts with $u = s$ and $dv = s \exp[-s^2/2]/(2\pi)^{1/2} ds$ shows that the variance is equal to

$$2\sigma^2 \left[-s \cdot \exp(-s^2/2)/(2\pi)^{1/2} \Big|_0^\infty + \int_0^\infty \exp[-s^2/2]/(2\pi)^{1/2} ds \right]$$

The first term is zero and the second is 1/2 since this is an integration of half of the standard normal density. Thus the variance is σ^2.

5.45 The uniform density is:

$$f(x) = \begin{cases} 1 & 0 < x < 1 \\ 0 & \text{elsewhere} \end{cases}$$

Thus, the distribution function is

$$F(x) = \begin{cases} 1 & x \geq 1 \\ x & 0 < x < 1 \\ 0 & x \leq 0 \end{cases}$$

5.47 Suppose Mr. Harris bids $(1 + x)c$. Then his expected profit is:

$$\begin{aligned} & 0P(\text{low bid } < (1+x)c) + xcP(\text{low bid } \geq (1+x)c) \\ = \quad & xc \int_{(1+x)c}^{2c} \frac{3}{4c} ds \;=\; 3xc[2c - (1+x)c]/4c \;=\; 3c(x - x^2)/4 \end{aligned}$$

Thus, his profit is maximum when $x = 1/2$. So his bid is 3/2 times his cost. Thus, he adds 50 percent to his cost estimate.

5.49 I_0/I_i is distributed log-normal with $\alpha = 2$, $\beta^2 = .01$, $\beta = .1$. Thus,

$$\begin{aligned} P(7 \leq I_0/I_i \leq 7.5) &= F((\ln(7.5) - 2)/.1) - F((\ln(7) - 2)/.1) \\ &= F(.149) - F(-.54) \\ &= .5592 - .2946 = .2646 \end{aligned}$$

5.51 (a) $P(\text{between 3.2 and 8.4}) = F((\ln(8.4) + 1)/2) - F((\ln(3.2) + 1)/2)$
$= F(1.564) - F(1.0816) = .9406 - .8599 = .0807$

(b) $P(\text{greater than 5}) = 1 - F((\ln(5) + 1)/2) = 1 - F(1.305) = 1 - .904 = .0960$

5.53 When $\alpha = 2$ and $\beta = 3$, $\Gamma(2) = 1$. So

$$f(x) = \begin{cases} xe^{-x/3}/9 & x > 0 \\ 0 & x \leq 0 \end{cases}$$

Thus,

$$P(X < 5) = \int_0^5 f(x)dx = \frac{1}{9}\int_0^5 xe^{-x/3}dx$$

Integrating by parts gives

$$\begin{aligned} -\frac{1}{3}xe^{-x/3}\Big|_0^5 + \frac{1}{3}\int_0^5 e^{-x/3}dx &= -\frac{5}{3}e^{-5/3} - e^{-x/3}\Big|_0^5 \\ &= 1 - \frac{8}{3}e^{-5/3} = 0.4963. \end{aligned}$$

5.55 (a) The probability that the supports will survive, if $\mu = 3.0$ and $\sigma^2 = .09$, is

$$\begin{aligned} P(\text{supports will survive}) &= 1 - F\left(\frac{\ln(33) - 3.0}{.30}\right) = 1 - F(1.655) \\ &= 1 - .9508 = .0492 \end{aligned}$$

(b) If $\mu = 4.0$ and $\sigma^2 = .25$, then

$$\begin{aligned} P(\text{supports will survive}) &= 1 - F\left(\frac{\ln(33) - 4.0}{.50}\right) = 1 - F(-1.007) \\ &= F(1.007) = .8430 \end{aligned}$$

5.57 We can ignore the constant in the density since it is always positive. Thus, we need to maximize $h(x) = x^{\alpha-1}e^{-x/\beta}$. Taking the derivative

$$h'(x) = (\alpha - 1)x^{\alpha-2}e^{-x/\beta} - x^{\alpha-1}e^{-x/\beta}/\beta = x^{\alpha-2}e^{-x/\beta}(\alpha - 1 - x/\beta)$$

Setting the derivative equal to zero gives the solution $x = \beta(\alpha-1)$. For $\alpha > 1$, the derivative is positive for $x < \beta(\alpha - 1)$ and negative for $x > \beta(\alpha - 1)$. Thus, $\beta(\alpha - 1)$ is a maximum. Note that $x = 0$ is a point of inflection when $\alpha > 2$. When $\alpha = 1$, $h(x) = e^{-x/\beta}$ which has a maximum in the interval $[0,\infty]$ at $x = 0$. When $0 < \alpha < 1$, the derivative does not vanish on $(0,\infty)$ and $h(x)$ is unbounded as x decreases to 0.

5.59 Since the number of breakdowns is a Poisson random variable with parameter $\lambda = .2$, the interval between breakdowns is an exponential random variable with parameter $\beta = 1/\lambda = 1/.2 = 5$.

(a) The probability that the interval is less than 1 week is $1 - e^{-1/5} = .1812$ or 18.12 percent.

(b) The probability that the interval is greater than 5 weeks is $e^{-5/5} = .3678$ or 36.78 percent.

5.61 Let N be a random variable having the Poisson distribution with parameter αt. Then $P(N=0) = (\alpha t)^0 e^{-\alpha t}/0! = e^{-\alpha t}$. Thus, $P(\text{waiting time is} > t) = e^{-\alpha t}$ and $P(\text{waiting time is } \le t) = 1 - e^{-\alpha t}$.

5.63 The beta density is

$$f(x) = \frac{\Gamma(\alpha+\beta)}{\Gamma(\alpha)\Gamma(\beta)} x^{\alpha-1}(1-x)^{\beta-1}$$

for $0 < x < 1$, $\alpha > 0$, and $\beta > 0$. For $\alpha = 3$ and $\beta = 3$

$$f(x) = \frac{\Gamma(6)}{\Gamma(3)\Gamma(3)} x^2(1-x)^2 = \frac{5!}{2!2!} x^2(1-x)^2 = 30(x^2 - 2x^3 + x^4)$$

Thus,

$$\begin{aligned}\int_0^1 f(x)dx &= 30\int_0^1 (x^2 - 2x^3 + x^4)dx = 30(x^3/3 - x^4/2 + x^5/5)\big|_0^1 \\ &= 30(1/3 - 1/2 + 1/5) = 1\end{aligned}$$

as required.

5.65 (a) The mean of the beta distribution is given by $\mu = \alpha/(\alpha+\beta)$. Thus, in the case where $\alpha = 1$ and $\beta = 4$, $\mu = 1/(1+4) = 1/5 = .2$

(b) When $\alpha = 1$ and $\beta = 4$, the beta density is

$$\frac{\Gamma(5)}{\Gamma(1)\Gamma(4)} x^0(1-x)^3 = \frac{4!}{0!3!}(1-x)^3 = 4(1-x)^3$$

Thus, the required probability is given by

$$4\int_{.25}^1 (1-x)^3 dx = -(1-x)^4\big|_{.25}^1 = (.75)^4 = .3164$$

5.67 Let X be Weibull random variable with $\alpha = .1$, $\beta = .5$ representing the battery lifetime. Then the density is $f(x) = (.1)(.5)x^{-.5}e^{-.1x^{.5}}$ for $x > 0$. Thus,

$$P(X \le 100) = \int_0^{100} (.1)(.5)x^{-.5}e^{-.1x^{.5}}dx$$

Using the change of variable $y = x^{.5}$ gives:

$$P(X \le 100) = .1\int_0^{10} e^{-.1y}dy = -e^{-.1y}\big|_0^{10} = 1 - e^{-1} = .6321$$

5.69 The probability is

$$\int_{4,000}^{\infty} (.025)(.500)x^{-.5}e^{-(.025)x^{.500}}\,dx = \int_{\sqrt{4,000}}^{\infty} .025e^{-.025y}\,dy = e^{-.025\sqrt{4,000}} = .2057$$

5.71 (a) The joint probability distribution of X_1 and X_2 is

$$f(x_1, x_2) = \frac{\binom{2}{x_1}\binom{1}{x_2}\binom{2}{2-x_1-x_2}}{\binom{5}{2}},$$

where $x_1 = 0, 1, 2$, $x_2 = 0, 1$, and $0 \le x_1 + x_2 \le 2$. The joint probability distribution $f(x_1, x_2)$ can be summarized in the following table:

	$f(x_1, x_2)$	X_2: 0	1	Total
X_1	0	.1	.2	.3
	1	.4	.2	.6
	2	.1	0	.1
	Total	.6	.4	1

(b) Let A be the event that $X_1 + X_2 = 0$ or 1, then

$$P(A) = f(0,0) + f(0,1) + f(1,0) = .1 + .2 + .4 = .7$$

(c) By (a), the marginal distribution of X_1 is

$$\begin{aligned} f_1(0) &= f(0,0) + f(0,1) = .1 + .2 = .3 \\ f_1(1) &= f(1,0) + f(1,1) = .4 + .2 = .6 \\ f_1(2) &= f(2,0) + f(2,1) = .1 + 0 = .1 \end{aligned}$$

(d) Since

$$f_2(0) = f(0,0) + f(1,0) + f(2,0) = .1 + .4 + .1 = .6,$$

the conditional probability distribution of X_1 given $X_2 = 0$ is

$$\begin{aligned} f_1(0|0) &= \frac{f(0,0)}{f_2(0)} = \frac{.1}{.6} = \frac{1}{6} \\ f_1(1|0) &= \frac{f(1,0)}{f_2(0)} = \frac{.4}{.6} = \frac{4}{6} \end{aligned}$$

$$f_1(2|0) = \frac{f(2,0)}{f_2(0)} = \frac{.1}{.6} = \frac{1}{6}$$

5.73 (a) $P(X_1 < 1, X_2 < 1) = F(1,1)$

$$= \int_0^1 \int_0^1 x_1 x_2 dx_2 dx_1 = \frac{1}{2}\int_0^1 x_1 dx_1 = \left.\frac{x_1^2}{4}\right|_0^1 = 1/4$$

(b) The probability that the sum is less than 1 is given by:

$$\int_0^1 \int_0^{1-x_1} x_1 x_2 dx_2 dx_1 = (1/2)\int_0^1 x_1(1-x_1)^2 dx_1$$
$$= (1/2)(x_1^4/4 - 2x_1^3/3 + x_1^2/2)\Big|_0^1 = (1/2)(1/4 - 2/3 + 1/2) = 1/24$$

5.75 The joint distribution function is given by:

$$F(x_1, x_2) = \int_0^{x_1} \int_0^{x_2} s_1 s_2 ds_2 ds_1 = \frac{1}{2}\int_0^{x_1} x_2^2 s_1 ds_1 = x_1^2 x_2^2/4$$

for $0 < x_1 < 2$ and $0 < x_2 < 1$. Thus, the distribution function is

$$F(x_1, x_2) = \begin{cases} 0 & x_1 \le 0 \text{ or } x_2 \le 0 \\ x_1^2 x_2^2/4 & 0 < x_1 < 2 \text{ and } 0 < x_2 < 1 \\ x_2^2 & x_1 \ge 2 \text{ and } 0 < x_2 < 1 \\ x_1^2/4 & 0 < x_1 < 2 \text{ and } x_2 \ge 1 \\ 1 & x_1 \ge 2 \text{ and } x_2 \ge 1 \end{cases}$$

The distribution function of X_1 is

$$F_1(x_1) = \int_0^{x_1} f_1(s_1) ds_1 = \int_0^{x_1} \frac{1}{2} s_1 ds_1 = x_1^2\ 4 \quad \text{for } 0 < x_1 < 2.$$

Thus,

$$F_1(x_1) = \begin{cases} 0 & x_1 \le 0 \\ x_1^2/4 & 0 < x_1 < 2 \\ 1 & x_1 \ge 2 \end{cases}$$

Similarly,

$$F_2(x_2) = \int_0^{x_2} f_2(s_2) ds_2 = \int_0^{x_2} 2 s_2 ds_2 = x_2^2 \quad \text{for } 0 < x_2 < 1.$$

Thus,

$$F_2(x_2) = \begin{cases} 0 & x_2 \le 0 \\ x_2^2 & 0 < x_2 < 1 \\ 1 & x_2 \ge 1 \end{cases}$$

It is easy to see that $F_1(x_1) \cdot F_2(x_2) = F(x_1, x_2)$. Thus, the random variables are independent.

5.77 The joint distribution function is given by

$$F(x,y) = \int_0^x \int_0^y \frac{6}{5}(u+v^2)dvdu = \frac{3x^2y}{5} + \frac{2xy^3}{5} \qquad \text{for } 0<x<1,\ 0<y<1$$

Thus, the joint distribution is

$$F(x,y) = \begin{cases} 0 & x \le 0 \text{ or } y \le 0 \\ (3/5)x^2y + (2/5)xy^3 & 0<x<1,\ 0<y<1 \\ (3/5)y + (2/5)y^3 & x \ge 1,\ 0<y<1 \\ (3/5)x^2 + (2/5)x & 0<x<1,\ y \ge 1 \\ 1 & x \ge 1,\ y \ge 1 \end{cases}$$

The probability of the region in the preceding exercise is given by

$$\begin{aligned} F(.5,.6) - F(.2,.6) - F(.5,.4) + F(.2,.4) &= .1332 - .03168 - .0728 + .01472 \\ &= .04344 \end{aligned}$$

5.79 (a) By definition

$$f_1(x|y) = \frac{f(x,y)}{f_2(y)} = \begin{cases} (x+y^2)/(\frac{1}{2}+y^2) & \text{for } 0<y<1,\ 0<x<1 \\ 0 & \text{elsewhere.} \end{cases}$$

(b) Thus,

$$f_1(x|.5) = \frac{x+.5^2}{\frac{1}{2}+.5^2} = \begin{cases} 4x/3 + 1/3 & \text{for } 0<x<1 \\ 0 & \text{elsewhere.} \end{cases}$$

(c) The mean is given by

$$\int_0^1 x(4x/3+1/3)dx = (4x^3/9 + x^2/6)\big|_0^1 = 11/18$$

5.81 (a) To find k, we must integrate the density and set it equal to 1. Thus,

$$\int_0^1 \int_0^2 \int_0^\infty k(x+y)e^{-z}dzdydx \int_0^1 \int_0^2 k(x+y)dydx$$

$$= \int_0^1 k(2x+2)dx = 3k = 1$$

Thus, $k = 1/3$.

(b) $P(X < Y,\ Z > 1)$

$$\begin{aligned} &= \frac{1}{3}\int_0^1\int_x^2\int_1^\infty (x+y)e^{-z}dzdydx \;=\; \frac{1}{3e}\int_0^1\int_x^2 (x+y)dydx \\ &= \frac{1}{3e}\int_0^1 (2x+2-3x^2/2)dx \;=\; 5/(6e) \;=\; .3066 \end{aligned}$$

5.83 (a) Notice that $f(x_1, x_2)$ can be factored into

$$\frac{1}{\sqrt{2\pi}\sigma}\exp\left[\frac{-1}{2\sigma^2}(x_1-\mu_1)^2\right]\cdot\frac{1}{\sqrt{2\pi}\sigma}\exp\left[\frac{-1}{2\sigma^2}(x_2-\mu_2)^2\right]$$

Thus, X_1 and X_2 are independent normal random variables. Thus,

$P(-8 < X_1 < 14, -9 < X_2 < 3) = P(-8 < X_1 < 14)P(-9 < X_2 < 3)$

$= (F((14-2)/10) - F((-8-2)/10))\,(F((3+2)/10) - F((-9+2)/10))$

$= (F(1.2) - F(-1))(F(.5) - F(-.7))$

$= (.8849 - .1587)(.6915 - .2420)$

$= (.7262)(.4495) = .3264$

(b) When $\mu_1 = \mu_2 = 0$ and $\sigma = 3$, the density is

$$f(x_1, x_2) = \frac{1}{2\pi\sigma}\exp\left[\frac{-1}{2\sigma^2}(x_1^2+x_2^2)\right]$$

Let R be the region between the two circles. We need to find

$$\int_R \frac{1}{2\pi\sigma^2}\exp\left[\frac{-1}{2\sigma^2}(x_1^2+x_2^2)\right]dx_1dx_2$$

Changing to polar coordinates gives

$$\begin{aligned} &\int_3^6\int_0^{2\pi}\frac{1}{2\pi\sigma^2}\exp(-r^2/(2\sigma^2))\boldsymbol{r}\,\boldsymbol{d\theta}\,\boldsymbol{dr} \\ &= \int_3^6 \frac{1}{\sigma^2}\exp(-r^2/(2\sigma^2))rdr \;=\; -\exp(-r^2/(2\sigma^2))\big|_3^6 \\ &= e^{-1/2} - e^{-2} \;=\; .4712 \end{aligned}$$

5.85 The expected value of $g(X_1, X_2)$ is

$$\begin{aligned} \int_{-\infty}^{\infty} g(x_1,x_2)f(x_1,x_2)dx_1dx_2 &= \int_0^1\int_0^2 (x_1+x_2)x_1x_2dx_2dx_1 \\ &= \int_0^1 (2x_1^2 + 8x_1/3)dx_1 \;=\; 2 \end{aligned}$$

5.87 The area of the rectangle is $X \cdot Y$. Thus the mean of the area distribution is given by

$$\int_{L-a/2}^{L+a/2} \int_{W-b/2}^{W+b/2} \frac{xy}{ab} dydx = \int_{L-a/2}^{L+a/2} \frac{x}{a} W dx = LW$$

The variance is given by

$$\begin{aligned}\int_{L-a/2}^{L+a/2} \int_{W-b/2}^{W+b/2} \frac{x^2y^2}{ab} dydx - (WL)^2 &= (W^2 + b^2/12)(L^2 + a^2/12) - (WL)^2 \\ &= \frac{1}{12}((aW)^2 + (bL)^2 + (ab)^2/12)\end{aligned}$$

5.89 (a) $E(X_1 + X_2) = E(X_1) + E(X_2) = 1 + (-1) = 0.$

(b) $Var(X_1 + X_2) = Var(X_1) + Var(X_2) = 5 + 5 = 10.$

5.91 (a) $E(X_1 + 2X_2 - 3) = E(X_1 + 2X_2) - 3 = E(X_1) + 2E(X_2) - 3 = 1 + 2(-2) - 3$
$= -6.$

(b) $Var(X_1 + 2X_2 - 3) = Var(X_1 + 2X_2) = Var(X_1) + 2^2 Var(X_2) = 3 + 2^2(5)$
$= 23.$

5.93 (a) $E(X_1 + X_2 + \cdots + X_{20}) = E(X_1) + E(X_2) + \cdots + E(X_{20})$
$= 20(10) = 200.$

(b) $Var(X_1 + X_2 + \cdots + X_{20}) = Var(X_1) + Var(X_2) + \cdots + Var(X_{20})$
$= 20(3) = 60.$

5.95 (a) The mgf is $M(t) = E[e^{tX}] = 0.25e^0 + 0.25(2)e^t + 0.25e^{2t}$

(b) Here $M'(t) =, = 0 + 0.25(2)e^t + 0.25(2)e^{2t}$ $E(X) = M'(0) = 0.5 + 0.5 = 1.0.$
Also, $M''(t) ,= (0.5)e^t + (0.25)4e^{2t}$ so $E(X^2) = M''(0) = 0.5 + 1.0 = 1.5.$

5.97 (a) For $t < 2$, the mgf is

$$M(t) = E[e^{tX}] = \int_0^\infty e^{tx}\, 2e^{-x} dx = \int_0^\infty e^{tx}\, 2e^{-2x} dx = \frac{2}{2-t} \int_0^\infty (2-t)e^{-(2-t)x} dx = \frac{2}{2-t}$$

(b) Here $M'(t) =, 2(2-t)^{-2}$ so $E(X) = M'(0) = \frac{1}{2}$ Also $M''(t) =, (2)(2)(2-t)^{-3}$ so
$E(X^2) = M''(0) = \frac{1}{2}$

5.99 Since X and Y are normal with the specified means and variances, the mgf of X is $M_X(t) = e^{2t+4t^2/2}$ and that of Y is $M_Y(t) = e^{t+9t^2i/2}$.

(a) By independence

$$M_{2X-3Y+5}(t) = E\left[e^{t(2X-3Y+5)}\right] = M_X(2t) \cdot M_Y(-3t)e^{t5}$$

Consequently,

$$M_{2X-3Y+5}(t) = e^{2(2t)+4(2t)^2/2} \cdot e^{(-3t)+9(-3t)^2/2} e^{5t} = e^{t(4-3+5)+(16+81)t^2/2}$$

which we recognize as the moment generating function of a normal random variable.

(b) The mean is the coefficient of t in the exponent and the variance is the coefficient of $t^2/2$ so the mean is 6 and the variance is 97.

5.101 (a) For any eleven observations the normal scores z_i, i=1, ..., 11, satisfy $F(z_i) = i/12$, thus using Table 3 the normal-scores are:

$$-1.38, -0.97, -0.67, -0.43, -0.21, 0, 0.21, 0.43, 0.67, 0.97, 1.38$$

(b) The observations on the times (sec.) between neutrinos are: .107, .196, .021, .283, .179, .854, .58, .19, 7.3, 1.18, 2.0. The normal-scores plot has a curved pattern that contradicts normality.

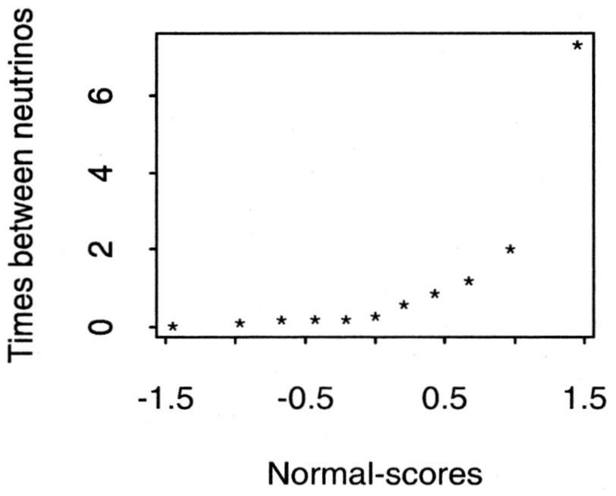

5.103 (a) The normal-scores plots of the logarithmic, square root and fourth root transformations for the decay time data are given in Figure 5.1. The log scale produces the straightest pattern.

(b) The normal-scores plots of the logarithmic, square root and fourth root transformations for the interarrival time data are given in Figure 5.2. The log scale produces the straightest pattern.

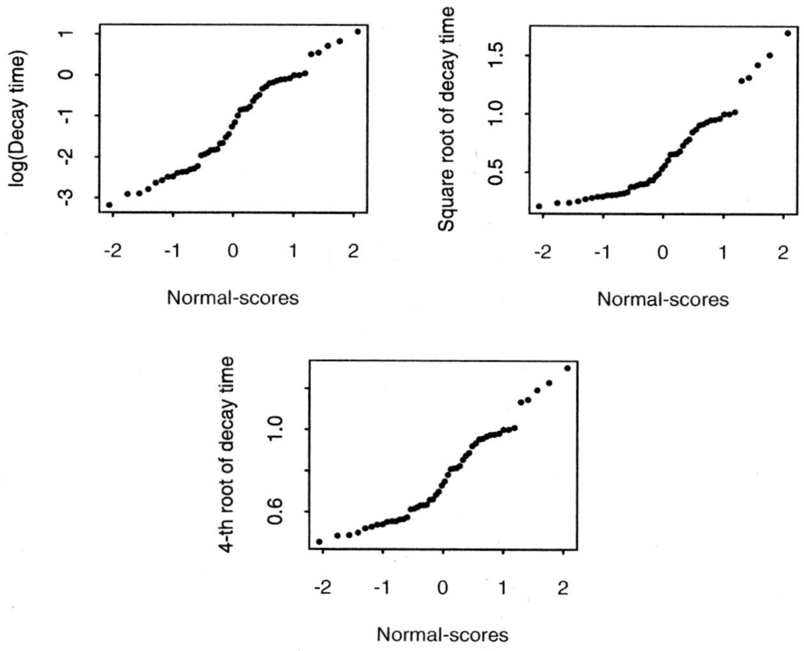

Figure 5.1: Normal-scores plots of the decay time data. Exercise 5.103a

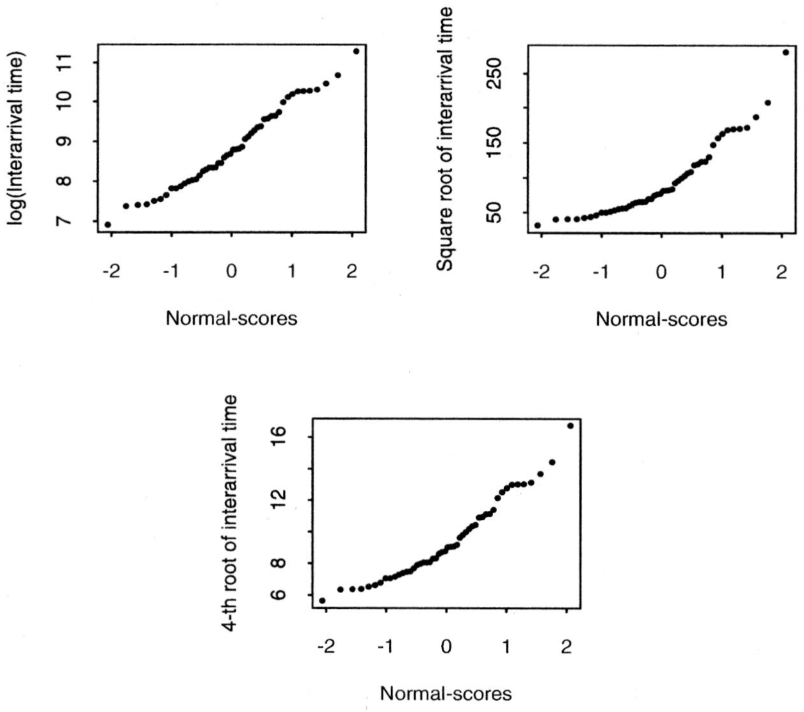

Figure 5.2: Normal-scores plots of the interarrival time data. Exercise 5.103b

5.105 (a) The density is

$$f(x) = \begin{cases} \alpha\beta x^{\beta-1}e^{-\alpha x^\beta} & \text{for } x > 0 \\ 0 & \text{elsewhere} \end{cases}$$

Thus, the corresponding distribution function is

$$F(x) = \begin{cases} \int_0^x \alpha\beta s^{\beta-1}e^{-\alpha s^\beta}ds = -e^{-\alpha s^\beta}\Big|_0^x = 1 - e^{-\alpha x^\beta} & \text{for } x > 0 \\ 0 & \text{elsewhere} \end{cases}$$

(b) We solve $u = F(x)$ for x. Since $u = F(x) = 1 - e^{-\alpha x^\beta}$, so $e^{-\alpha x^\beta} = 1 - u$ or $-\alpha x^\beta = \ln(1-u)$. The solution is then $x = (-(1/\alpha)\ln(1-u))^{1/\beta}$.

5.107 The eight exponential variates we generated, with $\beta = .2$, are:

```
0.053320   0.025485   0.071669   0.377028   0.208014   0.026851
   0.433405   0.747732
```

5.109 Let X be a random variable with density function

$$f(x) = \begin{cases} k(1-x^2) & \text{for } 0 < x < 1 \\ 0 & \text{elsewhere} \end{cases}$$

To find k we need to integrate $f(x)$ from 0 to 1 and set it equal to 1. Since,

$$\int_0^1 k(1-x^2)dx = k(x - x^3/3)\big|_0^1 = k(1-1/3) = k(2/3) = 1,$$

we have $k = 3/2$.

(a)

$$P(.1 < X < .2) = \frac{3}{2}\int_{.1}^{.2}(1-x^2)dx = \frac{3}{2}(x - x^3/3)\Big|_{.1}^{.2}$$

$$= \frac{3}{2}(.2 - .008/3 - .1 + .001/3) = \frac{3}{2}\frac{.293}{3} = \frac{.293}{2} = .1465$$

(b) $P(X > .5) = (3/2)(x - x^3/3)\big|_{.5}^{1} = (3/2)(1 - 1/3 - .5 + (.5^3)/3) = 5/16$

5.111 (a) P(error will be between -0.03 and 0.04)

$$= \int_{-.03}^{.04} 25dx = \int_{-.02}^{.02} 25dx = 1$$

(b) P(error will be between -0.005 and 0.005) $= 25(.01) = .25$

5.113 Let Z be a random variable with the standard normal distribution.

(a) $P(0 < Z < 2.5) = F(2.5) - .5 = .9938 - .5 = .4938$

(b) $P(1.22 < Z < 2.35) = F(2.35) - F(1.22) = .9906 - .8888 = .1018$

(c) $P(-1.33 < Z < -.33) = F(-.33) - F(-1.33)$

$= .3707 - .0918 = .2789$

(d) $P(-1.60 < Z < 1.80) = F(1.80) + F(-1.60) = .9641 - .0548 = .9093$

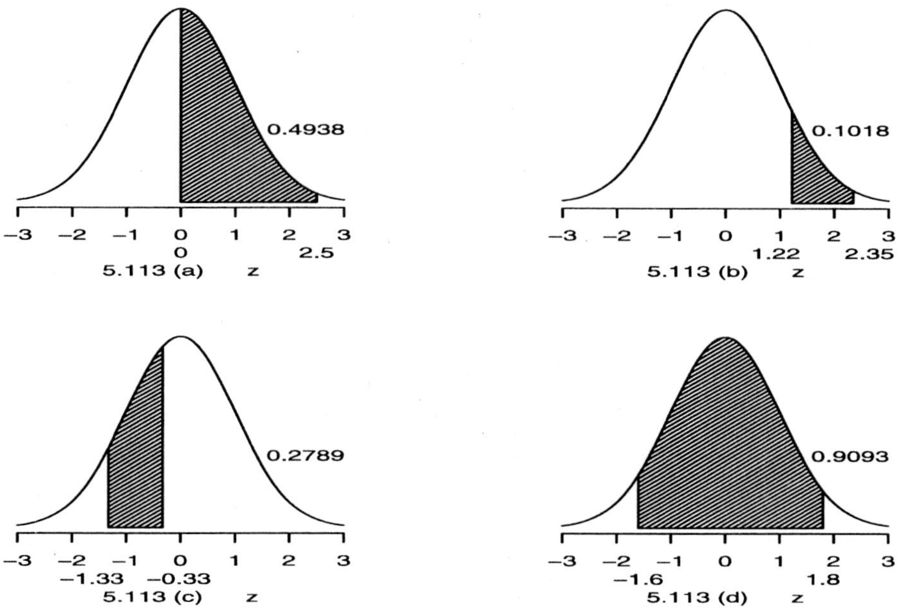

5.115 (a) $F(z_{.10}) = .90 = F(1.28)$, thus $z_{.10} = 1.28$

(b) $F(z_{.001}) = .999 = F(3.09)$, thus $z_{.001} = 3.09$

5.117 The density function is

$$f(x) = \begin{cases} \frac{1}{\sqrt{2\pi}\beta} x^{-1} \exp\left[-(\ln x - \alpha)^2/2\beta^2\right] & x > 0, \beta > 0 \\ 0 & \text{elsewhere} \end{cases}$$

(a)

$$\begin{aligned} P(X > 200) &= 1 - P(X < 200) = 1 - \int_0^{200} \frac{1}{\sqrt{2\pi}\beta} x^{-1} e^{-(\ln x - \alpha)^2/2\beta^2} dx \\ &= 1 - \int_{-\infty}^{\ln(200)} \frac{1}{\sqrt{2\pi}\beta} e^{-(y-\alpha)^2/2\beta^2} dy = 1 - F\left(\frac{\ln(200) - \alpha}{\beta}\right) \\ &= 1 - F\left(\frac{\ln(200) - 8.85}{1.03}\right) = 1 - F(-3.448) \\ &= F(3.448) = .9997 \end{aligned}$$

(b)

$$P(X < 300) = F(\frac{\ln(300) - 8.85}{1.03}) = F(-3.05) = .0011$$

5.119 The normal-scores plot of the suspended solids data casts suspicion on the normal assumption.

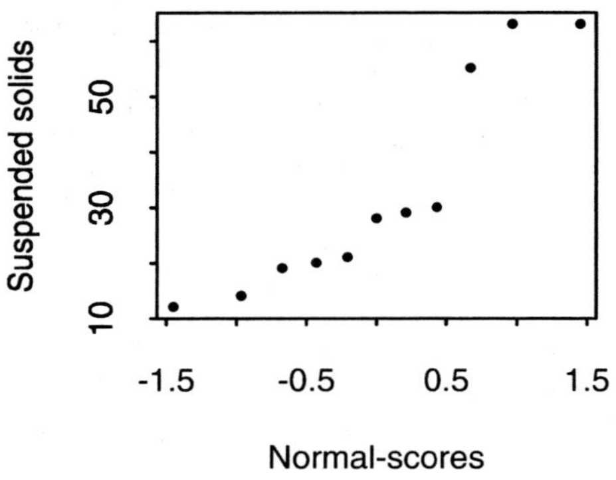

5.121 Let X be the gross sales volume, having the gamma distribution with $\alpha = 100\sqrt{n}$ and $\beta = \frac{1}{2}$ If the sales costs are 5,000 dollars per salesman then the expected profit is $E(P) = EX - 5n = \alpha\beta - 5n = 50\sqrt{n} - 5n$ which is maximized at $n = 25$.

5.123 Let (X, Y) have the joint density function

$$f(x, y) = \begin{cases} .04e^{-.2x-.2y} & \text{for } x > 0, y > 0 \\ 0 & \text{elsewhere} \end{cases}$$

(a) The marginal exponential distributions are identical and have mean $5.0 = 1/\ 0.2$.

$$f_1(x) = \int_0^\infty .04e^{-.2x-.2y}dy = .2e^{-.2x} \quad \text{for } x > 0$$

$$f_2(y) = \int_0^\infty .04e^{-.2x-.2y}dx = .2e^{-.2y} \quad \text{for } y > 0$$

(b)

$$\begin{aligned} &\int_0^\infty \int_0^\infty (x+y)(.04)e^{-.2x-.2y}dxdy \\ &= \int_0^\infty \int_0^\infty x(.04)e^{-.2x-.2y}dydx + \int_0^\infty \int_0^\infty y(.04)e^{-.2x-.2y}dxdy \\ &= \int_0^\infty x(.2)e^{-.2x}\left(\int_0^\infty .2e^{-.2y}dy\right)dx \end{aligned}$$

$$+ \int_0^\infty y(.2)e^{-.2y}\left(\int_0^\infty .2e^{-.2x}dx\right)dy$$
$$= \int_0^\infty x(.2)e^{-.2x}dx + \int_0^\infty y(.2)e^{-.2y}dy = \frac{1}{.2} + \frac{1}{.2} = 10.$$

(c) Here $E(X) = 1/.2 = E(Y)$, so $E(X+Y) = E(X) + E(Y) = 10$.

5.125 (a) $E(3X_1 + 5X_2 + 2) = E(3X_1 + 5X_2) + 2 = 3E(X_1) + 5E(X_2) + 2$
$= 3(-5) + 5(1) + 2 = -8$.

(b) $Var(3X_1 + 5X_2 + 2) = Var(3X_1 + 5X_2) = 3^2Var(X_1) + 5^2Var(X_2)$
$= 3^2(3) + 5^2(4) = 127$.

5.127 Let X be the maximum attenuation. We know that X has a normal distribution with mean 10.1 and standard deviation 2.8.

(a) $P(X < 6.0) = F(6.0 - 10.1)/2.7) = F(-1.519) = 0.0644$. The population maximum attenuation is modeled by the normal distribution so probability corresponds to proportion. About 6.44 % of the products have a maximum attenuation less than 6 dB.

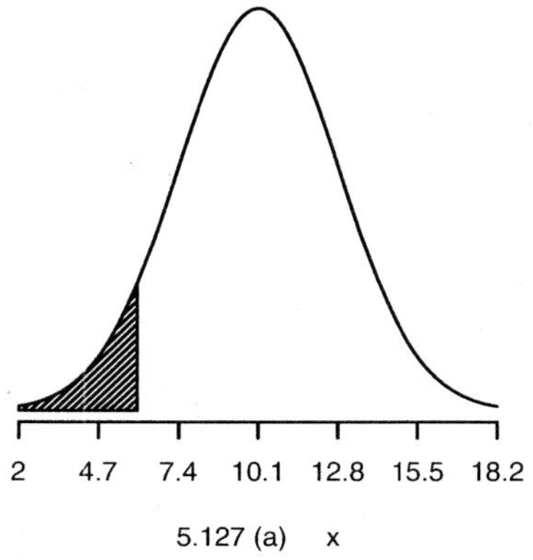

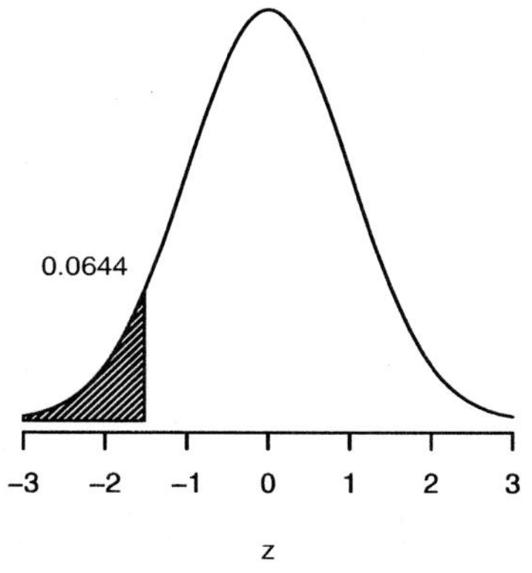

5.127 (a)

(b) $P(6 < X < 14) = F((14 - 10.1)/2.7) - F((6 - 10.1)/2.7) = F(1.444) - F(-1.518) = 0.9256 - 0.0644 = 0.8612$.

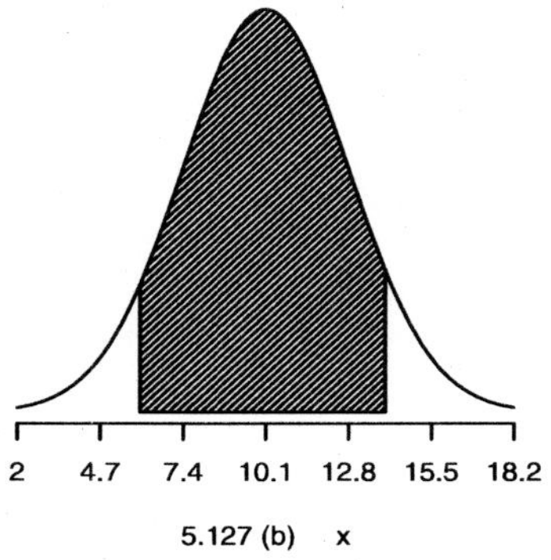

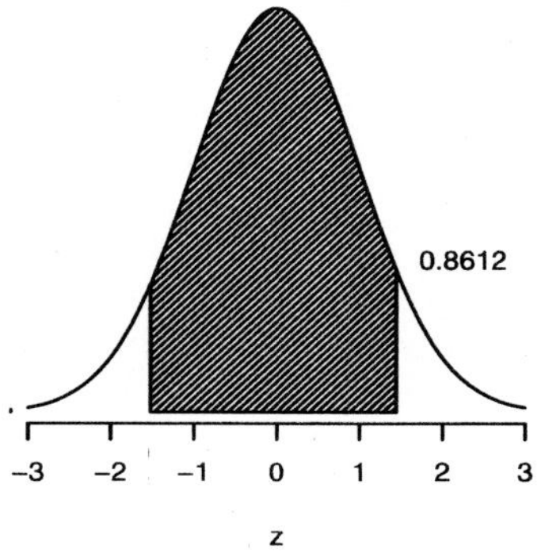

5.127 (b)

Chapter 6

SAMPLING DISTRIBUTIONS

6.1 Only if the pieces are put on the assembly line such that every 20-th piece is random with respect to the characteristic being measured. If twenty molds are dumping their contents, in sequential order, onto the assembly line then the sample would consist of output from a single mold . It would not be random.

6.3 (a) A typical member of the population does not vacation on a luxury cruise. The sample taken would be biased.

(b) This sample will very likely be biased. Those will high incomes will tend to respond while those with low incomes will tend to not respond.

(c) Everyone feels that unfair things should be stopped. The way the question is phrased biases the responses.

6.5 (a) The number of samples (given that order does not matter) is

$$\binom{7}{2} = \frac{7 \cdot 6}{2 \cdot 1} = 21.$$

(b) The number of samples (given that order does not matter) is

$$\binom{24}{2} = \frac{24 \cdot 23}{2 \cdot 1} = 276.$$

6.7 (a) The probability of each of the numbers is given in the table:

Number	−4	−3	−2	−1	0	1	2	3	4
Probability	1/6	2/15	1/10	1/15	1/15	1/15	1/10	2/15	1/6

The mean of the distribution is

$$\frac{1}{6}(-4) + \frac{2}{15}(-3) + \cdots + \frac{2}{15}(3) + \frac{1}{6}(4)$$
$$= \frac{1}{6}(4-4) + \frac{2}{15}(3-3) + \frac{1}{10}(2-2) + \frac{1}{15}(1-1) + \frac{1}{15}(0) = 0.$$

The variance is

$$\frac{2}{6}(16) + \frac{4}{15}(9) + \frac{2}{10}(4) + \frac{2}{15}(1) + \frac{1}{15}(0) = 8.667.$$

(b) The 50 samples are shown in Table 6.1 along with their means.

Table 6.1. 50 Samples of Size 10 Taken Without Replacement

obs.	obs.	obs.	obs.	obs.	obs.	obs.	obs.	obs.	obs.	mean
1	3	-1	4	4	4	-4	-4	-4	-2	0.1
-2	3	3	-2	2	-4	4	-4	-3	-4	-0.7
4	4	-2	-3	-3	2	1	3	4	0	1.0
3	3	-3	4	2	1	-2	0	4	0	1.2
4	0	-1	-2	3	1	2	4	-1	1	1.1
1	-2	-3	-4	2	4	-2	3	4	-4	-0.1
-1	2	1	-1	-4	-3	0	-4	3	1	-0.6
-4	-2	2	1	-4	-3	3	2	-2	1	-0.6
1	-3	2	-4	2	4	3	-4	-3	-2	-0.4
1	4	4	4	-4	3	-1	-2	2	-4	0.7
4	0	-1	1	-3	1	4	-3	4	2	0.9
-4	4	2	-4	2	-2	-2	-1	4	3	0.2
2	-1	-2	4	-4	0	-4	1	3	-2	-0.3
-4	-1	-3	4	0	-4	-3	1	2	4	-0.4
-2	4	4	-1	-4	1	4	1	-4	-3	0.0
-3	0	-3	-2	0	-4	4	-1	1	-2	-1.0
4	-4	3	4	-4	-1	3	2	0	1	0.8
3	3	-3	4	-3	1	3	-3	-4	-1	0.0
3	-4	-4	4	4	3	4	-3	2	-2	0.7
4	0	2	-3	-3	-3	-2	3	-4	-4	-1.0
2	-1	4	-3	2	3	4	4	-4	4	1.5
3	2	3	4	4	4	3	1	2	-2	2.4
4	-3	0	-1	-3	-3	0	-4	4	-4	-1.0
-3	3	-4	3	0	1	-4	-4	-3	-4	-1.5
-2	-4	-2	-3	4	4	3	-3	1	-3	-0.5
3	-1	-2	-3	1	-4	-2	3	4	2	0.1
4	2	-4	3	-2	2	-2	1	4	3	1.1
0	-2	4	-2	0	-1	2	3	-3	-4	-0.3
1	-4	4	4	-2	-3	-4	-4	0	3	-0.5
2	4	3	-2	3	1	0	0	-4	-3	0.4
4	0	2	-1	1	-4	-4	-4	-3	4	-0.5
4	-3	-1	-3	-3	0	4	4	3	1	0.6
-4	-1	0	-3	4	-3	2	1	3	4	0.3
0	4	-4	3	-2	2	0	4	2	-4	0.5

Table 6.1. (continued)

obs.	obs.	obs.	obs.	obs.	obs.	obs.	obs.	obs.	obs.	mean
3	2	-2	4	-4	4	3	4	0	-3	1.1
-4	2	0	-2	-3	2	-1	-2	-4	-2	-1.4
-3	-1	-2	4	2	4	0	1	2	2	0.9
2	-4	-2	-3	-1	3	1	-2	-4	-3	-1.3
-4	3	1	0	-1	0	-3	-3	2	4	-0.1
2	2	4	-4	-4	4	4	-3	-3	-2	0.0
-4	1	4	4	3	2	-2	-3	4	-3	0.6
0	-4	-4	0	-4	1	4	-3	1	4	-0.5
-4	4	-3	1	4	-1	-4	-3	-4	-2	-1.2
4	-3	-3	4	4	-3	1	2	3	-1	0.8
-2	3	2	-4	-1	3	-4	-1	-3	1	-0.6
4	2	4	-3	-4	-4	-1	-3	-4	-2	-1.1
-3	-1	1	2	-3	0	-4	-2	4	2	-0.4
0	-1	4	-4	2	-2	3	-3	-2	3	0.0
-4	2	1	-2	-2	-1	-3	-4	3	-3	-1.3
-4	4	4	4	-3	2	3	3	3	4	2.0

(c) The mean of the 50 sample means is .034. The sample variance of the 50 sample means is .8096.

(d) According to Theorem 6.1, the distribution of the means has mean 0 and variance

$$\frac{\sigma^2}{n} \cdot \frac{N-n}{N-1} = \frac{8.667}{10} \cdot \frac{20}{29} = .5977.$$

The sample values in part (c) compare well with these theoretical values.

6.9 A table of each outcome and the mean follows:

Outcome	Mean	Outcome	Mean
1, 1	1.0	3, 1	2.0
1, 2	1.5	3, 2	2.5
1, 3	2.0	3, 3	3.0
1, 4	2.5	3, 4	3.5
2, 1	1.5	4, 1	2.5
2, 2	2.0	4, 2	3.0
2, 3	2.5	4, 3	3.5
2, 4	3.0	4, 4	4.0

Thus, the distribution of these values is:

Value	No. of Ways Obtained	Probability
1.0	1	.0625
1.5	2	.1250
2.0	3	.1875
2.5	4	.2500
3.0	3	.1875
3.5	2	.1250
4.0	1	.0625

where the probability is (*no. ways*)$\times(.25)^2$. Consequently, the distribution of $\bar{X}$ has mean

$$\mu_{\bar{X}} = 1(.0625) + 1.5(.1250) + \cdots + 3.5(.1250) + 4(.0625) = 2.5$$

which may be obtained directly from the symmetry of the distribution. The distribution of $\bar{X}$ has variance

$$\begin{aligned}\sigma^2_{\bar{X}} &= (1-2.5)^2(.0625) + (1.5-2.5)^2(.1250) + \cdots \\ &\quad + (3.5-2.5)^2(.1250) + (4-2.5)^2(.0625) = .625\end{aligned}$$

Now the mean of the original distribution is also 2.5 and the variance is 1.25. Thus, Theorem 6.1 yields 2.5 and $1.25/2 = .625$ as the mean and the variance of the distribution of the sample mean of two observations. These agree exactly as they must.

6.11 The variance of the sample mean $\bar{X}$, based on a sample of size n, is σ^2/n. Thus the standard deviation, or standard error of the mean is $\sigma/\sqrt{n}$.

(a) The standard deviation for a sample of size 50 is $\sigma/\sqrt{50}$. The standard deviation for a sample is size 200 is $\sigma/\sqrt{200}$. That is, the ratio of standard errors is

$$\frac{\sigma/\sqrt{200}}{\sigma/\sqrt{50}} = \frac{\sqrt{50}}{\sqrt{200}} = \frac{1}{2},$$

so the standard error is halved.

(b) The ratio of standard errors is

$$\frac{\sigma/\sqrt{900}}{\sigma/\sqrt{400}} = \frac{\sqrt{400}}{\sqrt{900}} = \frac{2}{3},$$

so the standard error for a sample size 900 is 2/3rd's that for sample size 400.

(c) The standard error for a sample size 25 is

$$\frac{\sqrt{225}}{\sqrt{25}} = 3$$

times as large as that for sample size 225.

(d) The standard error for a sample size 40 is

$$\frac{\sqrt{640}}{\sqrt{40}} = 4$$

times as large as that for sample size 640.

6.13 We need to find $P(|\bar{X} - \mu| < .6745 \cdot \sigma/\sqrt{n})$. Since the standard deviation of the mean is $\sigma/\sqrt{n}$, the standardized variable $(\bar{X} - \mu)/(\sigma/\sqrt{n})$ is approximately a normal random variable for large n (central limit theorem). Thus , we need to find:

$$P(|\frac{\bar{X} - \mu}{\sigma/\sqrt{n}}| < .6745).$$

Now, interpolating in Table 3 gives

$$P(\frac{\bar{X} - \mu}{\sigma/\sqrt{n}} < .6745) = .75.$$

Thus,

$$P(\frac{\bar{X} - \mu}{\sigma/\sqrt{n}} \leq -.6745) = P(\frac{\bar{X} - \mu}{\sigma/\sqrt{n}} \geq .6745) = .25$$

so

$$P(|\frac{\bar{X} - \mu}{\sigma/\sqrt{n}}| < .6745) = .75 - .25 = .50.$$

The probability that the mean of a random sample of size n , from a population with standard deviation σ, will differ from μ by less than $(.6745)(\sigma/\sqrt{n})$ is approximately .5 for sufficiently large n.

6.15 Here $\sigma/\sqrt{n} = 0.0042/\sqrt{40} = 0.000664$ and

$$P(0.2245 \leq \overline{X} \leq 0.2260) = P(-.776 \leq \frac{\bar{X} - 0.2250}{.0.000664} \leq 1.506)$$

Since $n = 40$ is large , the central limit theorem yields the approximation

$$\begin{aligned} P(-.776 \leq \frac{\bar{X} - 0.2250}{0.0006664} \leq 1.506) &\approx F(1.506) - F(-.776) \\ &= .7151. \end{aligned}$$

6.17 We need to find

$$\begin{aligned} P(\sum_{i=1}^{36} X_i > 6,000) &= P(\bar{X} > 166.67) = P(\bar{X} - 163 > 3.67) \\ &= P(\frac{\bar{X} - 163}{18/6} > 1.222). \end{aligned}$$

Since $n = 36$ is relatively large, we use the central limit theorem to approximate this probability by

$$1 - F(1.222) = .111.$$

6.19 Let $\mu_X = E(X)$ be the expected value of X. First we will show that $E(X + Y) = E(X) + E(Y)$. Let

$f(x, y)$ be the joint density function of X and Y. Then, if X takes on discrete values x_i and Y takes on discrete values y_j,

$$\begin{aligned} E(X+Y) &= \sum_{i=0}^{\infty}\sum_{j=0}^{\infty}(x_i+y_j)f(x_i,y_j) \\ &= \sum_{i=0}^{\infty}x_i\sum_{j=0}^{\infty}f(x_i,y_j)+\sum_{j=0}^{\infty}y_j\sum_{i=0}^{\infty}f(x_i,y_j). \end{aligned}$$

But, $\sum_{j=0}^{\infty} f(x_i, y_j) = f_1(x)$, the marginal density of X. Similarly for Y. Thus

$$E(X+Y) = \sum_{i=0}^{\infty} x_i f_1(x_i) + \sum_{j=0}^{\infty} y_j f_2(y_j) = E(X) + E(Y)$$

by definition. By induction

$$E(\sum_{i=1}^{n} X_i) = \sum_{i=1}^{n} E(X_i).$$

Thus,

$$\begin{aligned} \mu_{\bar{X}} &= E(\frac{\sum_{i=1}^{n} X_i}{n}) \\ &= \frac{1}{n}E(\sum_{i=1}^{n} X_i) = \frac{1}{n}\sum_{i=1}^{n} E(X_i). \end{aligned}$$

But, each $E(X_i) = \mu$ so $\mu_{\bar{X}} = \mu$.

6.21 The mean of the data is $\bar{x} = 23$ and the sample standard deviation is 6.39. Thus, if the data are from a normal population with $\mu = 20$, the statistic

$$t = \frac{\bar{x}-\mu}{s/\sqrt{n}} = \frac{23-20}{6.39/\sqrt{6}} = 1.15$$

is the value of a t random variable with 5 degrees of freedom. The entry in Table 4 for $\alpha = 10$ and $\nu = 5$ is 1.476. Before the data are observed, we know that

$$P(\frac{\bar{X}-\mu}{S/\sqrt{n}} > 1.15) > .10.$$

The data do not give strong evidence against the ambulance service's claim.

6.23 The population of pitch is normal so that the χ^2 distribution with $n-1$ degrees of freedom applies.

$$P(S^2 > 0.122) = P(\frac{(n-1)S^2}{\sigma^2} > \frac{(n-1)0.122}{\sigma^2})$$

$$= P(\frac{(n-1)S^2}{\sigma^2} > \frac{(9)0.122}{0.065}) = P(\frac{9S^2}{\sigma^2} > 16.892).$$

with $n-1$ degrees of freedom. From Table 5, with $\nu = 9$ degrees of freedom, we see that

$$P(\frac{9S^2}{\sigma^2} > 18.307) = .05.$$

and a computer calculation shows that $P(9S^2\,/\,\sigma^2 > 16.892) = .0504$ Thus, the probability that the claim will be rejected even though $\sigma^2 = 0.065$ is about .05 or 5 percent.

6.25 We need to find

$$1 - P(\frac{1}{7} \le \frac{S_1^2}{S_2^2} \le 7).$$

Since the samples are independent and from normal populations, the statistic S_1^2/S_2^2 has an F distribution with $n_1 - 1 = n_2 - 1 = 7$ degrees of freedom for both the numerator and the denominator. From Table 6(b), we see that

$$P(\frac{S_1^2}{S_2^2} > 6.99) = .01.$$

Using the relation

$$F_{1-\alpha}(\nu_1, \nu_2) = \frac{1}{F_\alpha(\nu_2, \nu_1)},$$

we know that

$$P(\frac{S_1^2}{S_2^2} < \frac{1}{6.99}) = .01.$$

Thus, approximately

$$P(\frac{S_1^2}{S_2^2} < \frac{1}{7} \text{ or } \frac{S_1^2}{S_2^2} > 7) = P(\frac{S_1^2}{S_2^2} < \frac{1}{7}) + P(\frac{S_1^2}{S_2^2} > 7) = .01 + .01 + = .02.$$

6.27 We need to find

$$P(S^2 > 180) = P(\frac{(n-1)S^2}{\sigma^2} > \frac{4 \cdot 180}{144}) = P(\frac{(n-1)S^2}{\sigma^2} > 5).$$

Since the sample is from a normal population, the statistic $(n-1)S^2\,/\,\sigma^2$ has a χ^2 distribution with $n - 1 = 4$ degrees of freedom. The density of the χ^2 distribution with 4 degrees of freedom is

$$f(x) = \frac{1}{4} x e^{-x/2}, \qquad x > 0.$$

Thus,

$$P(S^2 > 180) = P(\frac{(n-1)S^2}{\sigma^2} > 5) = \int_5^\infty \frac{1}{4} x e^{-x/2}\, dx.$$

Integrating by parts with $u = x$ and $dv = e^{-x/2}$ gives

$$P(S^2 > 180) = -e^{-x/2}(\frac{1}{2}x + 1)\Big|_5^{\infty} = e^{-5/2}(\frac{5}{2} + 1) = e^{-5/2}\frac{7}{2} = .2873.$$

6.29 The probability that the ratio of the larger to the smaller sample variance exceeds 3 is

$$1 - P(\frac{1}{3} \le \frac{S_1^2}{S_2^2} \le 3) = 1 - \int_{1/3}^{3} \frac{6x}{(1+x)^4}\,dx.$$

Let $u = x + 1$ or $x = u - 1$. Then

$$\begin{aligned} P(\frac{1}{3} \le \frac{S_1^2}{S_2^2} \le 3) &= \int_{4/3}^{4} \frac{6(u-1)}{u^4}\,du = \frac{-3}{u^2}\Big|_{4/3}^{4} + \frac{2}{u^3}\Big|_{4/3}^{4} \\ &= \frac{24}{16} - \frac{52}{64} = .6875. \end{aligned}$$

Thus,

$$1 - P(\frac{1}{3} \le \frac{S_1^2}{S_2^2} \le 3) = 1 - .6875 = .3125.$$

6.31 First $Z_1 - Z_2$ has a normal distribution with mean 0 and variance 2 so $(Z_1 - Z_2)/\sqrt{2}$ has a standard normal distribution and is independent of $Z_3^2 + Z_3^4 + Z_5^2 + Z_6^2$ which is the sum of 4 independent standard normal variables squared and so has a chi square distribution with 4 degrees of freedom. According to the representation of t, the ratio

$$\frac{(Z_1 - Z_2)/\sqrt{2}}{\sqrt{\dfrac{Z_2^2 + Z_3^2 + Z_4^2 + Z_5^2}{4}}} = \frac{Z_1 - Z_2}{\sqrt{\dfrac{Z_2^2 + Z_3^2 + Z_4^2 + Z_5^2}{2}}}$$

has a t distribution with 4 degrees of freedom.

6.33 Since χ_1^2 and χ_2^2 are independent, they can be represented as

$$\chi_1^2 = \sum_{i=1}^{\nu_1} Z_i^2 \quad \text{and} \quad \chi_2^2 = \sum_{i=\nu_1+1}^{\nu_1+\nu_2} Z_i^2$$

where the Z_i's are independent standard normal random variables. Then

$$\chi_1^2 + \chi_2^2 = \sum_{i=1}^{\nu_1} Z_i^2 + \sum_{i=\nu_1+1}^{\nu_1+\nu_2} Z_i^2, = \sum_{i=1}^{\nu_1+\nu_2} Z_i^2$$

which is the sum of $\nu_1 + \nu_2$ squared standard normal variables. Consequently, it has the χ^2 distribution with $\nu_1 + \nu_2$ degrees of freedom.

6.35 We know that X_i has mgf $(1 - 2t)^{-2i}$ so that $\sum_{i=1}^{n} X_i$ has mgf

$$M_{\sum_{i=1}^{5} X_i}(t) = \prod_{i=1}^{5} \frac{1}{(1 - 2t)^{-2i}} = \frac{1}{(1 - 2t)^{-30}}$$

which is the mgf of a Gamma distribution with parameters $\alpha = 2$ and $\beta = 30$.

6.37 a) We know that X_i has mgf

$$M_i(t) = e^{t\mu_i + \frac{1}{2}t^2\sigma_i^2}$$

so, by independence,

$$M_{X_1-3X_2+2X_3-5}(t) = E[e^{tX_1 - 3tX_2+2tX_3-5}] = M_{X_1}(t)M_{X_2}(-3t)M_{X_3}(2t)e^{-5t}$$

This last product is equal to

$$e^{t\mu_1 + \frac{1}{2}t^2\sigma_1^2} \cdot e^{(-3t)\mu_2 + \frac{1}{2}(-3t)^2\sigma_2^2} \cdot e^{(2t)\mu_3 + \frac{1}{2}(2t)^2\sigma_3^2} \cdot e^{-5t}$$

$$e^{t(\mu_1 - 3\mu_2 + 2\mu_3 - 5) + \frac{1}{2}t^2(\sigma_1^2 + 9\sigma_2^2 + 4\sigma_3^2)}$$

This is the form of the moment generating function of a normal distribution.

(b) From the form of $M_{X_1-3X_2+2X_3-5}(t)$ we see that the mean is the coefficient of t or $\mu_1 - 3\mu_2 + 2\mu_3 - 5 = 2 - 3(1) + 2(-1) - 5 = -8$. The variance is the coefficient of $t^2/2$ or $\sigma_1^2 + 9\sigma_2^2 + 4\sigma_3^2 = 4 + 9(9) + 4(1) = 89$.

6.39 a) We know that X_i has mgf

$$M_i(t) = e^{t\mu_i + \frac{1}{2}t^2\sigma_i^2}$$

so, by independence,

$$M_{7X_1+X_2-2X_3+7}(t) = E[e^{t7X_1 + tX_2-2tX_3+7}] = M_{X_1}(7t)M_{X_2}(t)M_{X_3}(-2t)e^{7t}$$

This last product is equal to

$$e^{(7t)\mu_1 + \frac{1}{2}(7t)^2\sigma_1^2} \cdot e^{t\mu_2 + \frac{1}{2}(t)^2\sigma_2^2} \cdot e^{(-2t)\mu_3 + \frac{1}{2}(-2t)^2\sigma_3^2} e^{t7}$$

$$= e^{t(7\mu_1 + \mu_2 - 2\mu_3 - 7) + \frac{1}{2}t^2(49\sigma_1^2 + \sigma_2^2 + 4\sigma_3^2)}$$

This is the form of the moment generating function of a normal distribution.

(b) From the form of $M_{7X_1+X_2-2X_3-7}(t)$ we see that the mean is the coefficient of t or $7\mu_1 + \mu_2 - 2\mu_3 + 7 = 7(-4) + 0 - 2(3) + 7 = -27$. The variance is the coefficient of $t^2/2$ or $49\sigma_1^2 + \sigma_2^2 + 4\sigma_3^2 =$

$49(1) + 4 + 4(1) = 57.$

6.41 (a) We know from Exercise 6.40 that X_i has mgf

$$[pe^t / (1 - (1 - p) e^t]^{r_i} \quad \text{for} \quad pe^t < 1$$

so

$$M_{X_1+X_2+\cdots+X_n}(t) = M_{X_1}(t) M_{X_2}(t) \cdots M_{X_n}(t) \quad \text{by independence}$$

This last product is equal to

$$[pe^t / (1 - (1 - p) e^t]^{\sum_{i=0}^{n} r_i} \quad \text{for} \quad pe^t < 1$$

(b) Note that this is of the form of negative binomial mgf but with the number of successes given by $r = \sum_{i=0}^{n} r_i$. The sum of independent negative binomial random variables, all having the same success probability p, is again a negative binomial random variable.

6.43 Let $Y = Z^3$ where Z is standard normal with density function

$$f(z) = \frac{1}{\sqrt{2\pi}} e^{-z^2/2}$$

Then, since the transformation is monotone increasing,

$$G(y) = P(Y \leq y) = P(Z^3 \leq y) = P(Z \leq y^{1/3}) = F(y^{1/3})$$

Differentiating with respect to y, we obtain the density

$$g(y) = f(y^{1/3}) \frac{1}{3} | y^{-2/3} | + f(y^{1/3}) | \frac{1}{3} y^{-2/3} | = \frac{2}{3\sqrt{2\pi}} y^{-2/3} e^{y^{2/3}/2}$$

6.45 Let $Y = \ln(X)$ where X has the density function $f(x) = e^{-x}$ for $x > 0$. Since the transformation is monotone increasing, for any $-\infty < y < -\infty$ we have

$$G(y) = P(Y \leq y) = P(\ln(X) \leq y) = P(X \leq e^y) = F(e^y) = 1 - e^{-e^y}$$

Differentiating with respect to y, we obtain the density function $g(y) = e^y e^{-e^y}$.

6.47 Let $y = h(x) = -\ln(x)$ which is monotone decreasing on $0 < x < 1$ where $f(x) = 1$. Also, on the range, the inverse function $w(y) = e^{-y}$ so

$$g(y) = f(w(y)) | w'(y) | = 1 e^{-y} \quad \text{for} \quad 0 < y < \infty$$

and $g(y) = 0$ elsewhere. This is the exponential distribution with $\beta = 1$.

6.49 When X and Y are independent and each has the same gamma distribution,

$$f_{Y/X}(z) = \int_{-\infty}^{\infty} |x| f_X(x) f_Y(xz)\,dx = \int_0^{\infty} x \frac{1}{\Gamma(\alpha)} \frac{1}{\beta^\alpha} x^{\alpha-1} e^{-x/\beta} \frac{1}{\Gamma(\alpha)} \frac{1}{\beta^\alpha} (xz)^{\alpha-1} e^{-(xz)/\beta}\,dx$$

This last integral becomes

$$\frac{1}{\Gamma(\alpha)^2} \int_0^{\infty} z^{\alpha-1} \frac{1}{\beta^{2\alpha}} x^{2\alpha-2} e^{-x(1+z)/\beta}\,dx = \frac{1}{\Gamma(\alpha)^2} \int_0^{\infty} z^{\alpha-1} x^{2\alpha-1} \frac{1}{\beta^{2\alpha}} e^{-x(1+z)/\beta}\,dx$$

Under the change of variable $v = (1+z)x/\beta$, and multiplying and dividing by $\Gamma(2\alpha)$ and $(1+z)^{2\alpha}$ we obtain

$$\frac{1}{\Gamma(\alpha)^2} \Gamma(2\alpha) \frac{z^{\alpha-1}}{(1+z)^{2\alpha}} \int_0^{\infty} \frac{1}{\Gamma(2\alpha)} v^{2\alpha-1} e^{-v}\,dx = \frac{\Gamma(2\alpha)}{\Gamma(\alpha)^2} \frac{z^{\alpha-1}}{(1+z)^{2\alpha}} \quad \text{for } 0 < z < 1$$

6.51 (a) Number the states from 1 to 50 according to their alphabetical order. Using the last two columns of the second page of Table 7, starting at row 11, to select 10 numbers between 1 and 50 by discarding numbers outside of this range as well as any previously drawn number gives

17, 2, 12, 1, 44, 18, 33, 39, 40, 41.

These correspond to Kentucky, Alaska, Idaho, Alabama, Utah, Louisiana, North Carolina , Rhode Island, South Carolina, and South Dakota.

(b) No. High population states like California would have many participants but could send only two. A student from a small state would have a better chance than a student in California.

6.53 For a random sample, each outcome must have equal probability. The probabilities are :

(a) $\frac{1}{21} = .0476$.

(b) $\frac{1}{153} = .0065$.

6.55 (a) Here $\sigma/\sqrt{n} = 9/6 = 1.5$ and we will bound the one-tailed probability

$$P(\bar{X} > 66.75) = P(\bar{X} - 63 > 66.75 - 63).$$

Using Chebyshev's theorem for the two-tailed probability

$$\begin{aligned} P(|\bar{X} - 63| > 66.75 - 63) &= P(|\bar{X} - 63| > \frac{66.75 - 63}{9/6} \cdot 9/6) \\ &\leq (\frac{9/6}{66.75 - 63})^2 = .16. \end{aligned}$$

Thus, $P(\bar{X} > 66.75) \leq .16$.

(b) We need to find

$$P(\frac{\bar{X} - 63}{\sigma/\sqrt{n}} > \frac{66.75 - 63}{9/6}) = P(\frac{\bar{X} - 63}{\sigma/\sqrt{n}} > 2.5)$$

which by the central limit theorem is the much smaller probability

$$P(Z > 2.5) = 1 - F(2.5) = 1 - .9938 = .0062.$$

6.57 Since the population is normal, the distribution of $\bar{X}$ is exactly normal and we need to find

$$\begin{aligned} P(|\bar{X} - \mu| \leq .02) &= P(|\frac{\bar{X} - \mu}{.04/5}| \leq \frac{.02}{.04/5}) = F(2.5) - F(-2.5) \\ &= .9938 - .0062) = .9876. \end{aligned}$$

6.59 We need to find

$$P(\frac{S_1^2}{S_2^2} > 4.0).$$

Since the samples are independent and from normal populations, the statistic S_1^2/S_2^2 has an F distribution with $n_1 - 1 = 8$ and $n_2 - 1 = 15$ degrees of freedom for the numerator and the denominator, respectively . From Table 6(b), we see that

$$P(\frac{S_1^2}{S_2^2} > 4.0) = .01,$$

so the probability is .01 or 1 percent.

6.61 The variance of the sample mean $\bar{X}$, based on a sample of size n, is σ^2/n. Thus the standard deviation, or standard error, of the mean is $\sigma/\sqrt{n}$.

(a) The standard deviation for a sample of size 100 is $\sigma/\sqrt{100}$. The standard deviation for a sample is size 200 is $\sigma/\sqrt{200}$. That is, the ratio of standard errors is

$$\frac{\sigma/\sqrt{200}}{\sigma/\sqrt{100}} = \frac{\sqrt{100}}{\sqrt{200}} = \frac{1}{\sqrt{2}} = .707,$$

so the standard error for sample size 200 is .707 times the standard error for sample size 100.

(b) The ratio of standard errors is

$$\frac{\sigma/\sqrt{300}}{\sigma/\sqrt{200}} = \frac{\sqrt{200}}{\sqrt{300}} = \sqrt{\frac{2}{3}} = .816,$$

so the standard error for a sample size 300 is .816 times that for sample size 200.

(c) The ratio of standard errors is

$$\frac{\sigma/\sqrt{90}}{\sigma/\sqrt{360}} = 2,$$

so the standard error doubles.

6.63 (a) The production process may not be stabilized when the first shaft is made each morning. Alternatively, the first shaft could be given extra attention and may not be representative.

(b) Buses and large trucks would take longer to pass the fixed point and thus would be more likely to be included in the sample than a small car.

Chapter 7

INFERENCES CONCERNING A MEAN

7.1 The error bound at the 95 percent level is given by

$$E = z_{.025}\sqrt{s^2\,/\,n} = (1.96)\cdot\sqrt{\frac{671330.9}{30}} = 293.2$$

Using $t_{0.025} = 2.045$ gives 305.9

7.3 Since the sample is fairly large, the error is, with approximately 95 percent confidence, less than or equal to

$$z_{.025}\frac{s}{\sqrt{n}} = 1.96\cdot\frac{1.250}{\sqrt{52}} = .340$$

7.5 Since the sample is fairly large, the error is, with approximately 98 percent confidence, less than or equal to

$$z_{.01}\frac{s}{\sqrt{n}} = 2.33\cdot\frac{3.057}{\sqrt{45}} = 1.062$$

7.7 Since the sample is fairly large, the error is, with approximately 95 percent confidence, less than or equal to

$$z_{.025}\frac{s}{\sqrt{n}} = 1.96\cdot\frac{14,054}{\sqrt{50}} = 3895.6$$

7.9 We need to find $z_{\alpha/2}$ such that

$$10 = z_{\alpha/2}\cdot(62.35)/\sqrt{80}.$$

Thus,

$$z_{\alpha/2} = \sqrt{80}\cdot 10/62.35 = 1.43.$$

From Table 3, we see that F(1.43) = .9236. Thus, $\alpha = 2(1 - .9236) = .1528$. The confidence level on the error bound of \$ 10 is about 84.7 percent.

7.11 Since $P\left(|X - \mu| \le z_{.005}\sigma/\sqrt{n}\right) = .99$, we need to choose n such that

$$z_{.005}\frac{\sigma}{\sqrt{n}} = .25.$$

In this case, $\sigma = 1.40$, $z_{.005} = 2.575$. Thus,

$$n = \left(\frac{(2.575)\cdot 1.40}{.25}\right)^2 = 207.9 \simeq 208.$$

7.13 The sample size $n = 9$ is small so we use $t_{.025} = 2.306$ for 8 degrees of freedom. We first calculate $\overline{x} = 1.334$ and $s = 0.674$ Since the population is normal, the 95 percent confidence interval for the mean product volume is given by

$$\bar{x} \pm t_{.025}\cdot\frac{s}{\sqrt{n}} = 1.334 \pm 2.306\cdot\frac{0.674}{\sqrt{9}} = 1.334 \pm 0.518$$

or from 0.816 to 1.852. We are 95 % confident that the mean product volume is between .816 and 1.852 gal.

7.15 The sample size $n = 9$ is small so we use $t_{.025} = 2.306$ for 8 degrees of freedom. We first calculate $\overline{x} = 114.00$ and $s = 8.34$ Since the population is normal, the 95 percent confidence interval for the key performance indicator is given by

$$\bar{x} \pm t_{.025}\cdot\frac{s}{\sqrt{n}} = 114.00 \pm 2.306\cdot\frac{8.34}{\sqrt{9}} = 114.00 \pm 6.41$$

or 107.59 to 120.41. We are 95 % confident that the mean of the key performance indicator is between 107.59 and 120.41.

7.17 Since the sample size $n = 30$ is large, we can use the large sample confidence interval with $\overline{x} = 1908.8$, $s = 327.1$, $n = 30$. Since $z_{.025} = 1.96$, the 95 percent confidence interval is

$$1908.8 - (1.96)\frac{327.1}{\sqrt{30}} < \mu < 1908.8 + (1.96)\frac{327.1}{\sqrt{30}}$$

or, $1791.7 < \mu < 2025.8$. We are 95% confident that the mean strength is between 1792 and 2026 pounds per square inch.

7.19 Since the sample size is small, if we can reasonably regard the data as a sample from a normal population, we can use the small sample confidence interval for μ with $\bar{x} = 30.91$, $s = .788$, $n = 10$, and $t_{.025}$ with 9 degrees of freedom, which is equal to 2.262. Thus, the 95 percent confidence interval is

$$30.91 - (2.262)\frac{.788}{\sqrt{10}} < \mu < 30.91 + (2.262)\frac{.788}{\sqrt{10}}$$

or, $30.35 < \mu < 31.47$. We are 95% confident that the mean thickness of paper is between 30.35 mm and 31.47 mm.

7.21 We are given $n = 36$, $\overline{x} = 3.5$ and $s = 0.8$.

(a) The sample size is large so the 90 % confidence interval is

$$\bar{x} \pm z_{.05} \cdot \frac{s}{\sqrt{n}} = 3.5 \pm 1.645 \cdot \frac{0.8}{\sqrt{36}} = 3.5 \pm 0.219$$

or 3.28 to 3.72. We are 090 % confident that the mean freshness is between 3.28 and 3.72.

(b) The population mean μ for all customers is unknown so we never know if it is covered by a particular confidence interval.

(c) Before we sample, the probability is .90 that the interval will cover μ. By the long run relative frequency interpretation of probability, if we take many different samples and calculate a 90 % confidence interval for each, about 90 % of the time they will cover μ.

7.23 We are given $n = 12$, $\overline{x} = 7.2$ and $s = 1.2$.

(a) The sample size is small so we assume that the population is normal. We use $t_{.025} = 2.201$ for 11 degrees of freedom. The 95 % confidence interval is

$$\bar{x} \pm t_{.025} \cdot \frac{s}{\sqrt{n}} = 7.2 \pm 2.201 \cdot \frac{1.2}{\sqrt{12}} = 7.2 \pm .76$$

or 6.4 to 8.0. We are 95 % confident that the mean stamping pressure is between 6.44 and 7.96 thousand psi.

(b) The population mean μ of maximum pressures from all possible occasions, is unknown so we never know if it is covered by a particular confidence interval.

(c) The sample size was small so we assumed the population was normal.

(d) Before we sample, the probability is .95 that the interval will cover μ. By the long run relative frequency interpretation of probability, if we take many different samples and calculate a 95% confidence interval for each, about 95 % of the time they will cover μ.

7.25 As discussed in Chapter 6, if a random sample of size n is taken from a population of size N, having mean μ and variance σ^2, then $\bar{X}$ has mean μ and variance $\sigma^2(N-n)/n(N-1)$. Thus, the formula for E becomes

$$E = z_{\alpha/2}\frac{\sigma}{\sqrt{n}} \cdot \sqrt{\frac{N-n}{N-1}}.$$

(a) In this case, $z_{.025} = 1.96, \quad s = 85, \quad N = 420,$ and $n = 50$. Thus,

$$E = 1.96 \cdot \frac{85}{\sqrt{50}}\sqrt{\frac{420-50}{420-1}} = 22.14.$$

Thus, we have 95 percent confidence that the error will be less than 22.14.

(b) In this case, $z_{.005} = 2.575, \quad \sigma = 12.2, \quad n = 40, \quad N = 200$. Thus,

$$E = 2.575\frac{12.2}{\sqrt{40}} \cdot \sqrt{\frac{200-40}{200-1}} = 4.454.$$

Thus, we have 99 percent confidence that the error will be less than 4.454.

7.27 Let X have a binomial distribution with a probability of success p.

(a) X/n is an unbiased estimator of p since the expected value of X/n is

$$\begin{aligned} E\left(\frac{X}{n}\right) &= \sum_{x=0}^{n}\frac{x}{n}\binom{n}{x}p^x(1-p)^{n-x} = \frac{p}{n}\sum_{x=1}^{n}x\binom{n}{x}p^{x-1}(1-p)^{n-x} \\ &= \frac{p}{n}\sum_{x=1}^{n}\frac{n!x}{x!(n-x)!}p^{x-1}(1-p)^{n-x} = \frac{p}{n}n\sum_{x=1}^{n}\frac{(n-1)!}{(x-1)!(n-x)!}p^{x-1}(1-p)^{n-x} \\ &= p\sum_{x=0}^{n-1}\frac{(n-1)!}{x!(n-1-x)!}p^x(1-p)^{n-1-x} = p \end{aligned}$$

(b) $(X+1)/(n+2)$ is not an unbiased estimator of p since

$$E\left(\frac{X+1}{n+2}\right) = \frac{1}{n+2}E(X+1) = \frac{1}{n+2}[E(X)+1] = \frac{np+1}{n+2} \neq p$$

7.29 (a) The MINITAB output is

	N	MEAN	STDEV	SE MEAN	95.0 PERCENT C.I.
C1	5	22.93	2.35	1.05	(20.02, 25.85)
C2	5	21.52	3.59	1.61	(17.06, 25.98)
C3	5	17.082	2.170	0.971	(14.386, 19.778)
C4	5	19.53	6.33	2.83	(11.66, 27.39)
C5	5	19.37	4.07	1.82	(14.31, 24.42)
C6	5	21.56	2.31	1.03	(18.69, 24.43)
C7	5	21.56	2.73	1.22	(18.16, 24.95)
C8	5	17.17	2.25	1.01	(14.37, 19.96)
C9	5	20.21	4.83	2.16	(14.21, 26.21)
C10	5	23.72	5.30	2.37	(17.15, 30.30)
C11	5	19.70	2.35	1.05	(16.78, 22.63)
C12	5	22.55	7.10	3.18	(13.73, 31.37)
C13	5	21.03	4.49	2.01	(15.45, 26.61)
C14	5	18.30	8.29	3.71	(8.00, 28.60)
C15	5	18.70	4.02	1.80	(13.70, 23.69)
C16	5	20.09	5.11	2.29	(13.74, 26.44)

C17	5	17.70	2.50	1.12	(14.60,	20.81)
C18	5	19.76	7.04	3.15	(11.01,	28.51)
C19	5	18.18	3.67	1.64	(13.63,	22.73)
C20	5	18.15	6.46	2.89	(10.13,	26.17)

The proportion of the 20 intervals that cover the true mean $\mu = 20$ is 1.00.

(b) The MINITAB output is

	N	MEAN	STDEV	SE MEAN	95.0 PERCENT C.I.	
C1	5	22.93	2.35	1.05	(20.02,	25.85)
C2	5	21.52	3.59	1.61	(17.06,	25.98)
C3	5	17.082	2.170	0.971	(14.386,	19.778)
C4	5	19.53	6.33	2.83	(11.66,	27.39)
C5	5	19.37	4.07	1.82	(14.31,	24.42)
C6	5	21.56	2.31	1.03	(18.69,	24.43)
C7	5	21.56	2.73	1.22	(18.16,	24.95)
C8	5	17.17	2.25	1.01	(14.37,	19.96)
C9	5	20.21	4.83	2.16	(14.21,	26.21)
C10	5	23.72	5.30	2.37	(17.15,	30.30)
C11	5	19.70	2.35	1.05	(16.78,	22.63)
C12	5	22.55	7.10	3.18	(13.73,	31.37)
C13	5	21.03	4.49	2.01	(15.45,	26.61)
C14	5	18.30	8.29	3.71	(8.00,	28.60)
C15	5	18.70	4.02	1.80	(13.70,	23.69)
C16	5	20.09	5.11	2.29	(13.74,	26.44)
C17	5	17.70	2.50	1.12	(14.60,	20.81)
C18	5	19.76	7.04	3.15	(11.01,	28.51)
C19	5	18.18	3.67	1.64	(13.63,	22.73)
C20	5	18.15	6.46	2.89	(10.13,	26.17)

The proportion of the 20 intervals that cover the true mean $\mu = 20$ is .90.

7.31 The data are: 48 out of 60 transceivers passed inspection.

(a) According to the example on page 117, the maximum likelihood estimator is

$$\widehat{p} = \overline{x} = \frac{\text{number of successes}}{\text{number of trials}} \quad .$$

The maximum likelihood estimate of the probability of passing inspection is $48/60 = .80$

(b) By independence, $P(\text{ next two transceivers will pass }) = p^2$. According to the invariance property of the maximum likelihood estimator on page 219, the maximum likelihood estimator of $P(\text{ next two transceivers will pass }) = p^2$ is $\widehat{p^2} = \widehat{p}^2 = (.8)^2 = .64$.

7.33 The data have mean $\overline{x} = 1.5$ calls dropped.

(a) According to the example on page 118, the maximum likelihood estimator is $\widehat{\lambda} = \overline{x}$. The maximum likelihood estimate of λ is 1.5 calls dropped.

(b) According to the invariance property of the maximum likelihood estimator on page 219, the maximum likelihood estimator of

$$P(\text{ no drops in next two calls }) = P(\text{ no drops in next call }) \times P(\text{ no drops in second call })$$

$$= e^{-\lambda} \times e^{-\lambda} = e^{-2\lambda}$$

is $$\widehat{P(\text{ no drops in next two calls })} = \widehat{e^{-2\lambda}} = e^{-2\widehat{\lambda}} = e^{-2(1.5)} = 0.05$$

7.35 The summary statistics are mean $\overline{x} = 114.00$ and $\sum_{i=1}^{9} (x_i - \overline{x})^2 = 556$

(a) According to the example on page 120, the maximum likelihood estimators are $\widehat{\mu} = \overline{x}$ and $\widehat{\sigma} = \sqrt{\sum_{i=1}^{n} (x_i - \overline{x})^2 / n}$. The maximum likelihood estimates are

$$\widehat{\mu} = \overline{x} = 114.00 \quad \text{and} \quad \widehat{\sigma} = \sqrt{\frac{\sum_{i=1}^{n} (x_i - \overline{x})^2}{n}}, = \sqrt{\frac{556}{9}} = 7.860$$

(b) According to the invariance property of page 219, the maximum likelihood estimator of σ / μ is

$$\widehat{\left(\frac{\sigma}{\mu}\right)} = \frac{\widehat{\sigma}}{\widehat{\mu}} = \frac{7.860}{114.00} = 0.0689$$

7.37 (a) In the case of the exponential distribution, the likelihood is

$$L(\beta) = \beta^{-n} e^{-\sum x_i / \beta}$$

Differentiating with respect to β,

$$L'(\beta) = \beta^{-n-1} e^{-\sum x_i / \beta} \left(-n + \frac{\sum x_i}{\beta} \right)$$

so, we see that $\hat{\beta} = \bar{X}$ is the maximum likelihood estimator.

(b) According to the invariance property of the maximum likelihood estimator on page 219, the maximum likelihood estimator of

$$P(X > 1) = 1 - P(X \leq 1) = e^{-1/\beta}$$

is $$\widehat{P(X > 1)} = e^{-1/\widehat{\beta}} = e^{-1/\overline{X}}$$

7.39 (a) The manufacturer wants to establish that mean time, μ, to set up a computer is less than 2 hours so that becomes the alternative hypothesis.

$$H_0 : \mu = 2 \qquad H_1 : \mu < 2$$

(b) If $\mu = 1.9$ the alternative hypothesis holds and the only possible error is failing to reject H_0. That is, the mean time to set up a computer is less than 2 hours but we fail to reject the null hypothesis that it is 2 hours.

(c) If $\mu = 2.0$ the null hypothesis holds and the only possible error is to reject H_0. That is, the mean time to set up a computer is 2 hours but we conclude that it is less than 2 hours.

7.41 (a) We want to show that the mean flying time, μ, is different from 56 minutes so that becomes the alternative hypothesis.

$$H_0 : \mu = 56 \qquad H_1 : \mu \neq 56$$

(b) If $\mu = 50$ the alternative hypothesis holds and the only possible error is failing to reject H_0. That is, the mean flying time is different from 56 but we fail to reject the null hypothesis that it is 56.

(c) If $\mu = 56$ the null hypothesis holds and the only possible error is to reject H_0. That is, the mean flying time is 56 minutes but we conclude that it is different from 56 minutes.

7.43 (a) We should make the engineers prove that the bridge is safe so that should be the alternative hypothesis. The null hypothesis then asserts it is unsafe. These hypothesis would need to be translated into a statement about a parameter, perhaps the probability of failure in next three years.

(b) The value $\alpha = .05$, or 1 in 20 chances of saying a unsafe bridge is safe, is too high for me. Even $\alpha = .01$, or 1 in 100 bridges, seems high when talking about personal safety.

7.45 The firm commits a Type I error if it erroneously rejects the null hypothesis that the dam is safe. If it erroneously accepts the null hypothesis that the dam is safe, it commits a Type II error.

7.47 We can assume from past experience that the standard deviation of the drying times is 2.4 minutes. The null hypothesis is that the mean $\mu = 20$. We reject the null hypothesis if $\bar{X} > 20.50$ minutes.

(a) the probability of a Type I error is the probability that $\bar{X} > 20.50$ when $\mu = 20$. Using a normal approximation to the distribution of the sample mean, this probability is given by

$$1 - F\left(\frac{20.50 - 20}{2.4/\sqrt{36}}\right) = 1 - F(1.25) = 1 - .8944 = .1056$$

(b) The probability of a Type II error when $\mu = 21$ is the probability that $\bar{X} < 20.50$ when $\mu = 21$. Using a normal approximation to the distribution of the sample mean, this probability is given by

$$F\left(\frac{20.50 - 20}{2.4/\sqrt{36}}\right) = F(-1.25) = .1056.$$

7.49 Using the normal approximation for the distribution of the sample mean, we need to find c such that

$$F\left(\frac{c - 100}{12/\sqrt{40}}\right) = .01.$$

Using the normal table, we see that

$$\frac{c - 100}{12/\sqrt{40}} = -2.33$$

so, we reject the null hypothesis if $\bar{x} > c$ where

$$c = 95.58.$$

7.51 (a) In this case, use the two-sided alternative $\mu \neq \mu_0$ where μ is the true mean daily inventory under the new marketing policy, and μ_0 is the true mean daily inventory under the old policy (Note: $\mu_0 = 1250$).

(b) The burden of proof is on the new policy. Thus, the alternative is $\mu < \mu_0$.

(c) The burden of proof is on the old policy. Thus, the alternative is $\mu > \mu_0$.

7.53 This is the large sample setting.

(a) 1. *Null hypothesis* $H_0 : \mu = 2000$

Alternative hypothesis $H_1 : \mu < 2000$

2. *Level of significance:* $\alpha = 0.05$.

3. *Criterion:* Using a normal approximation for the distribution of the sample mean, we reject the null hypothesis when

$$Z = \frac{\bar{X} - \mu_0}{s/\sqrt{n}} < -z_\alpha.$$

Since $\alpha = .05$ and $z_{.05} = 1.645$, the null hypothesis must be rejected if $Z < -1.645$.

4. *Calculations:* $\mu_0 = 2000$, $\bar{x} = 1798.4$, $s = 819.35$ and $n = 30$ so

$$Z = \frac{1798.4 - 2000}{819.35/\sqrt{30}} = -1.348$$

5. *Decision:* Because $-1.348 > -1.645$, the null hypothesis that $\mu = 2000$ is not rejected at level .05. The P-value $= P[Z < -1.348] = .089$, also shown in the figure, confirms that the evidence against the null hypothesis, $\mu = 2000$, is very weak.

(b) We could have failed to reject the null hypothesis when the mean amount of salt actually used is less than 2000 tons.

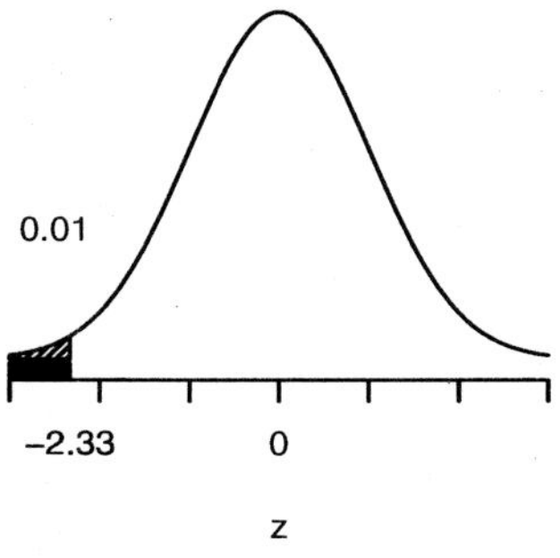

(a) Rejection Region

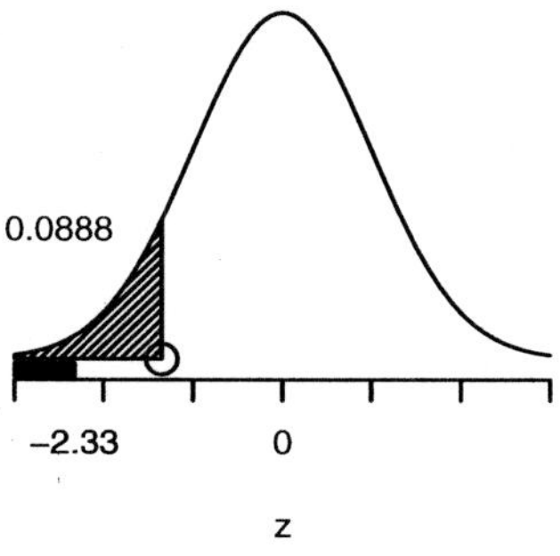

(b) P-value for Exercise 7.53

7.55 This is the large sample setting.

(a) The alternative hypothesis is the result we intend to establish.

1. *Null hypothesis* $H_0 : \mu = 3.6$
   Alternative hypothesis $H_1 : \mu < 3.6$

2. *Level of significance:* $\alpha = 0.025$.

3. *Criterion:* Using a normal approximation for the distribution of the sample mean, we reject the null hypothesis when
$$Z = \frac{\bar{X} - \mu_0}{s/\sqrt{n}} \quad < \quad - z_\alpha .$$
   Since $z_{.025} = 1.96$, the null hypothesis must be rejected if $Z < -1.96$.

4. *Calculations:* The observed $\bar{x} = 2.467$, $s = 3.057$ and $n = 45$. Since $\mu_0 = 3.6$
$$Z = \frac{2.467 - 3.6}{3.057/\sqrt{45}} = -2.49$$

5. *Decision:* Because $-2.49 < -1.96$, we reject the null hypothesis that $\mu = 3.6$ at level .025. The P-value $= P[Z < -2.40] = .006$ confirms that the evidence against the null hypothesis, $\mu = 3.6$, is strong.

(b) We could have rejected the null hypothesis that the mean number of unremovable defects is 3.6 and falsely concluded that it is less.

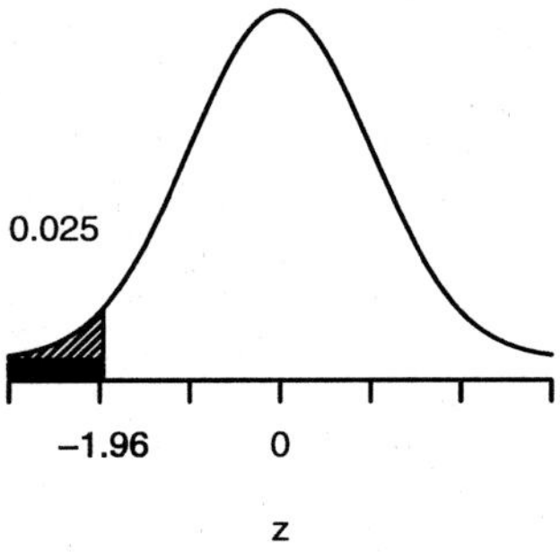

(a) Rejection Region

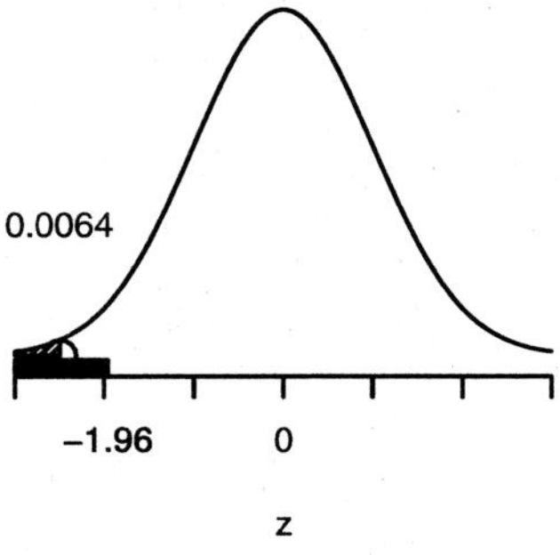

(b) P-value for Exercise 7.55

7.57 This is the small sample setting.

(a) The alternative hypothesis is the result we intend to establish concerning the key performance indicator.

1. *Null hypothesis* $H_0 : \mu = 107$

   Alternative hypothesis $H_1 : \mu \neq 107$

2. *Level of significance:* $\alpha = 0.05$.

3. *Criterion:* The population is normal and the sample size is small so we use the test statistic

$$t = \frac{\bar{X} - \mu_0}{s/\sqrt{n}}$$

   Since $t_{\alpha/2} = t_{.025} = 2.306$, for $n - 1 = 9 - 1 = 8$ degrees of freedom, the null hypothesis must be rejected if $t < -2.306$ or $t > 2.306$.

4. *Calculations:* $\mu_0 = 107$ and we find that $\bar{x} = 114.0$ and $s = 8.34$ so

$$t = \frac{114.0 - 107}{8.34/\sqrt{9}} = 2.52$$

5. *Decision:* Because $2.52 > 2.306$, we reject the null hypothesis that $\mu = 107$ at level .025. The P-value $= P[t < -2.52] + P[t > 2.52] = .036$ confirms that the evidence against the null hypothesis, $\mu = 107$, is somewhat strong.

(b) We could have rejected the null hypothesis that the mean key performance indicator is 107 and falsely concluded that it is different from 107.

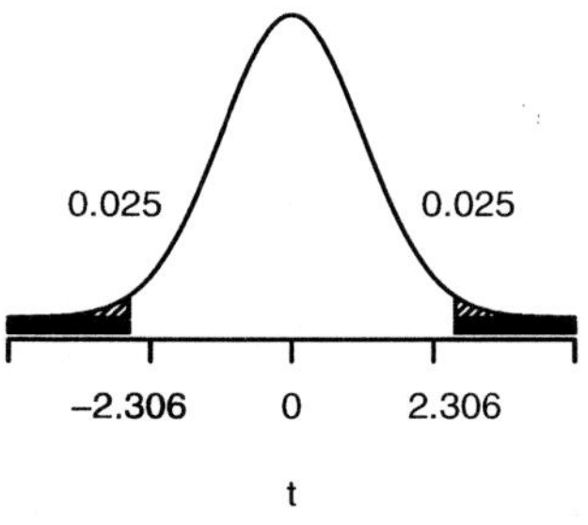

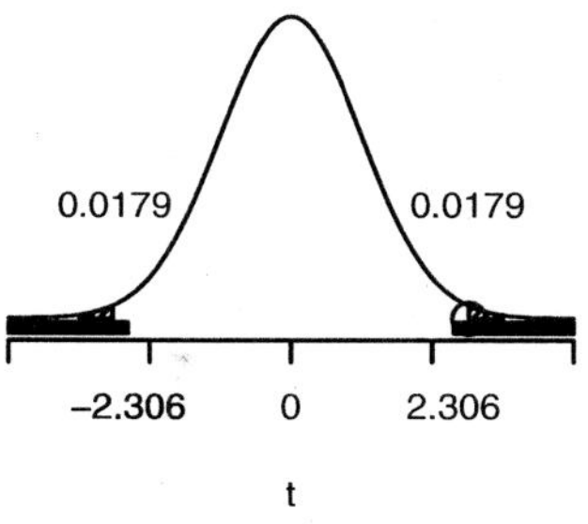

7.59 1. *Null hypothesis* $H_0 : \mu = 1.3$

Alternative hypothesis $H_1 : \mu > 1.3$

2. *Level of significance:* $\alpha = 0.05$.

3. *Criterion:* We Use the large sample normal approximation for the distribution of the sample mean and reject the null hypothesis when

$$Z = \frac{\bar{X} - \mu_0}{S/\sqrt{n}} > z_\alpha .$$

Since $\alpha = .05$ and $z_{.05} = 1.645$, the null hypothesis must be rejected if $Z > 1.645$.

4. *Calculations:* $\mu_0 = 1.3$, $\bar{x} = 1.4707$, $s = 0.5235$, and $n = 35$

$$Z = \frac{1.4707 - 1.3}{0.5235/\sqrt{29}} = 1.76$$

5. *Decision:* Because $1.76 > 1.645$, the null hypothesis that $\mu = 1.3$ is rejected. at level .05. The P-value $= P[Z > 1.76] = .039$ as shown in the figure. The evidence against the null hypothesis, $\mu = 1.3$, is moderately strong.

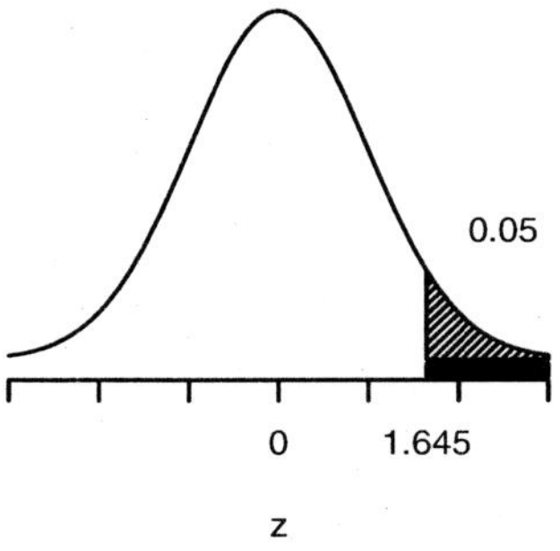

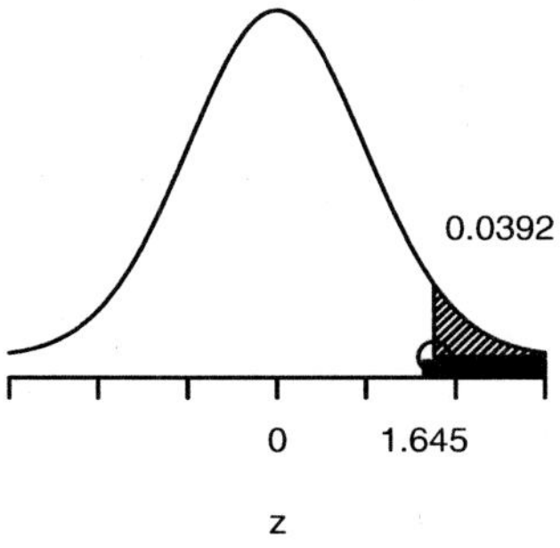

7.61 1. *Null hypothesis* $H_0 : \mu = 30.0$
Alternative hypothesis $H_1 : \mu \neq 30.0$

2. *Level of significance:* $\alpha = 0.05$.

3. *Criterion:* Since the sample is small, we can not use the normal approximation. If it is reasonable to assume that the data are from a distribution that is nearly normal, we can use the t statistic

$$t = \frac{\bar{X} - \mu_0}{S/\sqrt{n}}$$

Since the alternative hypothesis is two-sided, the critical region is defined by $t < -t_{.025}$ or $t > t_{.025}$ where $t_{.025}$ with 9 degrees of freedom is 2.262.

4. *Calculations:* $\mu_0 = 30.0$, $\bar{x} = 30.91$, $s = .778$, and $n = 10$ so

$$t = \frac{30.91 - 30.0}{.788/\sqrt{10}} = 3.652$$

5. *Decision:* Because $3.652 > 2.262$, we reject the null hypothesis at the .05 level of significance and conclude that the mean thickness of paper is different from 30.0mm. A computer calculation gives the The P-value $= P[t < -3.652] + P[t > 3.652] = .0054$ in the figure.

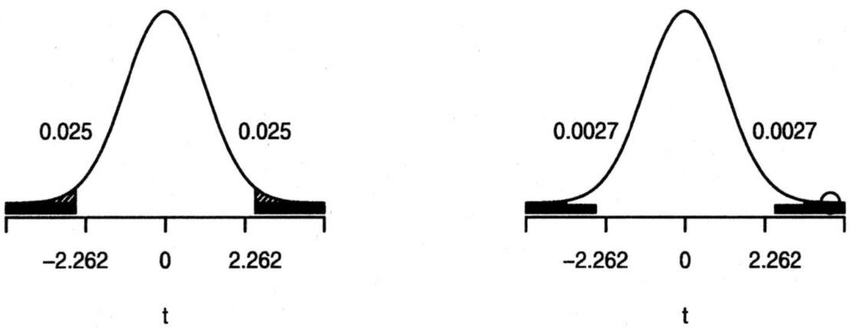

(a) Rejection region (b) P-value for Exercise 7.61

7.63 1. *Null hypothesis* $H_0 : \mu = 14.0$
Alternative hypothesis $H_1 : \mu \neq 14.0$

2. *Level of significance:* $\alpha = 0.05$.

3. *Criterion:* Assuming the population is normal, we can use the t statistic.

$$t = \frac{\bar{X} - \mu_0}{S/\sqrt{n}}$$

Since the alternative hypothesis is two-sided, the critical region is defined by $t < -t_{.025}$ or $t > t_{.025}$ where $t_{.025}$ with 4 degrees of freedom is 2.776.

4. *Calculations:* In this case, $\mu_0 = 14.0$, $n = 5$, $\bar{x} = 14.4$ and $s = .158$ so

$$t = \frac{14.4 - 14.0}{.158/\sqrt{5}} = 5.66.$$

5. *Decision:* Because $5.66 > 2.776$, we reject the null hypothesis in favor of the alternative hypothesis $\mu \neq 14.0$ at the .05 level of significance. From the t−table, the P-value is less than .005. A computer program gives the P-value 0.0048 in the figure.

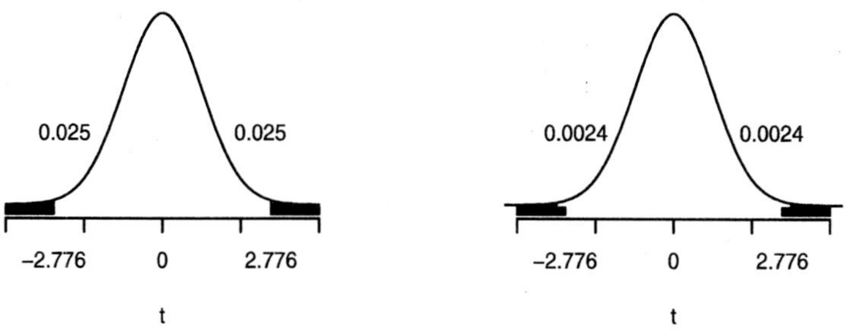

(a) Rejection region (b) P-value for Exercise 7.63

7.65 (a) `One-Sample T: Height(nm)`

```
Test of mu = 295 vs not = 295

Variable      N    Mean   StDev  SE Mean      T      P
Height(nm)   50  305.58   36.97     5.23   2.02  0.048
```

Since the P-value of the test is less than .05, we reject the null hypothesis at the .05 level of significance.

(b) `TEST MU=16.0 C2`

```
TEST OF MU = 16.00 VS MU N.E. 16.00

      N    MEAN   STDEV   SE MEAN      T   P VALUE
C2   11   27.55   10.02      3.02   3.82    0.0034
```

Since the P-value of the test is less than .05, we reject the null hypothesis at the .05 level of significance. The evidence against H_0 is very strong.

7.67 The MINITAB computer output for the Z test is

```
 N     Mean  SE Mean         95% CI
50   305.58     5.23  (295.33, 315.83)
```

(a) The value 320 nm is outside of the 95 % confidence interval. Consquently, at level $\alpha = 0.05$, we reject the null hypothesis $H_0 : \mu = 320$ in favor of the two-sided alternative.

(b) The value 310 nm lies inside the 95 % confidence interval. Consquently, at level $\alpha = 0.05$, we fail to reject the null hypothesis $H_0 : \mu = 320$.

(c) If $\alpha = 0.02$ the confidence interval would be centered at the same value but would be even wider. Therefore the value 310 nm also lies inside the 98 % confidence interval. At level $\alpha = 0.02$, we fail to reject the null hypothesis $H_0 : \mu = 310$.

7.69 (a) The value 1.6 lies inside the 90 % confidence interval. Consquently, at level $\alpha = 0.10$, we fail to reject the null hypothesis $H_0 : \mu = 1.5$.

(b) The value 2.2 is outside of the 90 % confidence interval. Consquently, at level $\alpha = 0.10$, we reject the null hypothesis $H_0 : \mu = 1.5$ in favor of the two-sided alternative.

(c) If $\alpha = 0.05$ the confidence interval would be centered at the same value but would be even wider. Therefore the value 1.6 also lies inside the 95 % confidence interval. At level $\alpha = 0.05$, we fail to reject the null hypothesis $H_0 : \mu = 1.6$.

7.71 (a) Here $\alpha = 0.03$ and $z_{0.03} = 1.88$. The power of a right-tailed test at $\mu_1 = 77$ is

$$\gamma(\mu_1) = P\left(Z > z_\alpha + \sqrt{n}\,\frac{(\,u_0 - \mu_1\,)}{\sigma} \right)$$

$$\text{where} \qquad z_\alpha + \sqrt{n}\,\frac{(\,\mu_0 - \mu_1\,)}{\sigma} = 1.88 + \sqrt{15}\,\frac{(\,75.2 - 77.0\,)}{3.6} = -0.0565$$

so the power is $\gamma(77.0) = P(Z > -.0565\,) = 0.523$.

(b) Here $\alpha = 0.05$ and $z_{0.05} = 1.645$ $\mu_1 = 77$. We calculate

$$-z_{\alpha/2} + \sqrt{n}\,\frac{(\,\mu_0 - \mu_1\,)}{\sigma} = -1.96 + \sqrt{15}\,\frac{(\,75.2 - 77.0\,)}{3.6} = -3.896$$

$$z_{\alpha/2} + \sqrt{n}\,\frac{(\,\mu_0 - \mu_1\,)}{\sigma} = 1.96 + \sqrt{15}\,\frac{(\,75.2 - 77.0\,)}{3.6} = 0.235$$

so the power is $\gamma(77.0) = P(\,Z < -3.886\,) + P(Z > 0.235\,) = 0.4906$.

7.73 Using MINITAB, with $n = 15, \alpha = 0.03$ and $s = 3.6$, we obtain the power $\gamma(77) = 0.522$. More conservatively, we could have chosen **1-Sample t**.

(a)

```
             Sample
Difference     Size     Power
       1.8       15  0.522209
```

(b) With $H_0 : \mu \neq 75.2$, we obtain the power $\gamma(77) = 0.491$.

```
             Sample
Difference     Size     Power
       1.8       15  0.490686
```

7.75 We entered .1:2.5/.1 in differences to obtain the values of power.

(a), (b) The power curve and OC curve are

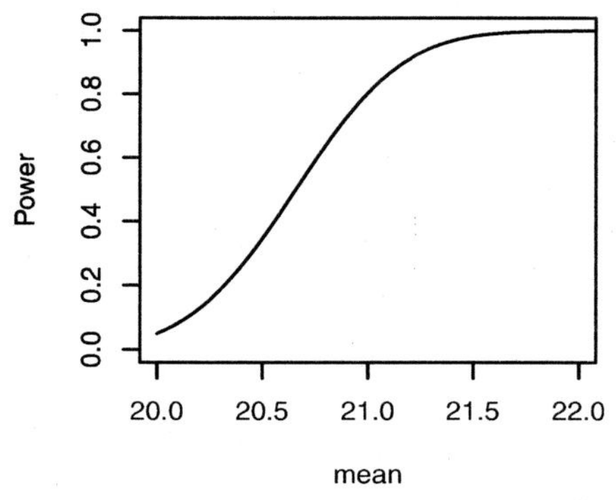

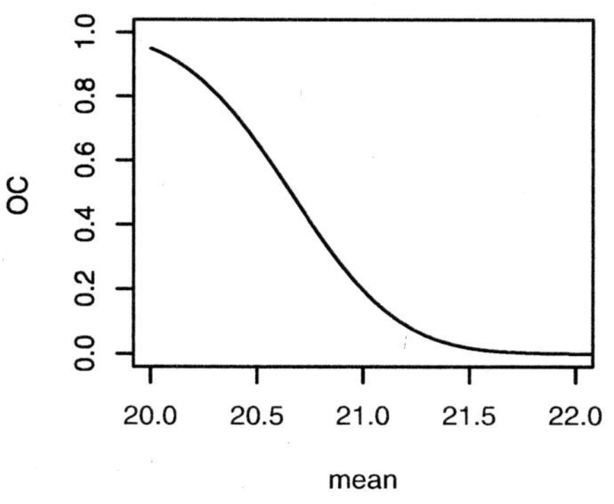

(a) Power Curve (b) Operating Characteristic for Exercise 7.75

7.77 The 95% large sample confidence interval for the mean strength of the aluminum alloy 1 is

$$\overline{x} \pm z_{.05/2}\frac{s}{\sqrt{n}} = 70.6966 \pm 1.96\frac{1.793}{\sqrt{58}} = 70.6966 \pm .4625$$

or $70.234 < \mu < 71.159$. We are 95 % confident that the mean strength is between 70.23 and 71.16 thousand psi.

7.79 The 95 % confidence interval is given by

$$\overline{x} \pm z_{\alpha/2}\frac{s}{\sqrt{n}} = 26.40 \pm 1.48,$$

so the interval is from 24.92 to 27.88. We are 95 % confident that the mean increase in pulse rate is between 24.92 and 27.88 beats per minute.

7.81 Since $t_{.025}$ with 11 degrees of freedom equals 2.201, the 95% confidence interval for the actual mean eccentricity of the can shafts is

$$\bar{x} \pm t_{.025}\frac{s}{\sqrt{n}} = 1.020 \pm 2.201\frac{.044}{\sqrt{12}} = 1.020 \pm .0279$$

or $.992 < \mu < 1.048$. We are 95 % confident that the mean eccentricity is between .992 and 1.048 inches.

7.83 First, we use the error bound from the normal distribution to get an initial estimate of the required sample size. Thus, we need to find n_1 such that

$$\frac{(14,380)(1.96)}{\sqrt{n_1}} = 10,000$$

(since $z_{.025} = 1.96$). Thus, $n_1 = 7.94 \simeq 8$. Now we use $t_{.025} = 2.365$ with 7 degrees of freedom to estimate n_2. Thus n_2 is given by

$$\frac{(14,380)(2.365)}{\sqrt{n_2}} = 10,000$$

or, $n_2 = 11.56 \simeq 12$. Next, we use $t_{.025} = 2.201$ with 11 degrees of freedom to estimate n_3. Thus n_3 is given by

$$\frac{(14,380)(2.201)}{\sqrt{n_3}} = 10,000$$

or, $n_3 = 10.017 \simeq 11$. Now, we use $t_{.025}$ with 10 degrees of freedom to find n_4. Proceeding as before, $n_4 = 10.265 \simeq 11$. Since we have converged to a sample of size 11, 11 observations would be required to have 95 percent confidence that the error is less than 10,000.

7.85 (a) The critical region is given by $Z > z_{.03} = 1.881$ where

$$Z = \frac{\overline{X} - 20}{2.4/\sqrt{50}}$$

Solving for $\overline{X}$ gives the critical region

$$\overline{X} > \frac{(1.881)(2.4)}{\sqrt{50}} + 20 = 20.6384.$$

(b) The calculation is shown below.

μ	$z = (20.6384 - \mu)/(2.4/\sqrt{50})$	F(z)
19.50	3.354	0.9996
19.75	2.617	0.9956
20.00	1.881	0.9700
20.25	1.144	0.8737
20.50	0.408	0.6582
20.75	−0.329	0.3711
21.00	−1.065	0.1433
21.25	−1.802	0.0358
21.50	−2.539	0.0056
21.75	−3.275	0.0005
22.00	−4.012	0.0000

A plot of the OC curve for the test with sample size 50 is given in Figure 7.1 (a). The OC curve for both tests are shown in Figure 7.1(b) where the dotted line corresponds to the test for sample size 50.

7.87 We are given $n = 180$, $\overline{x} = 1.65$ hours and $s = 0.82$ hours.

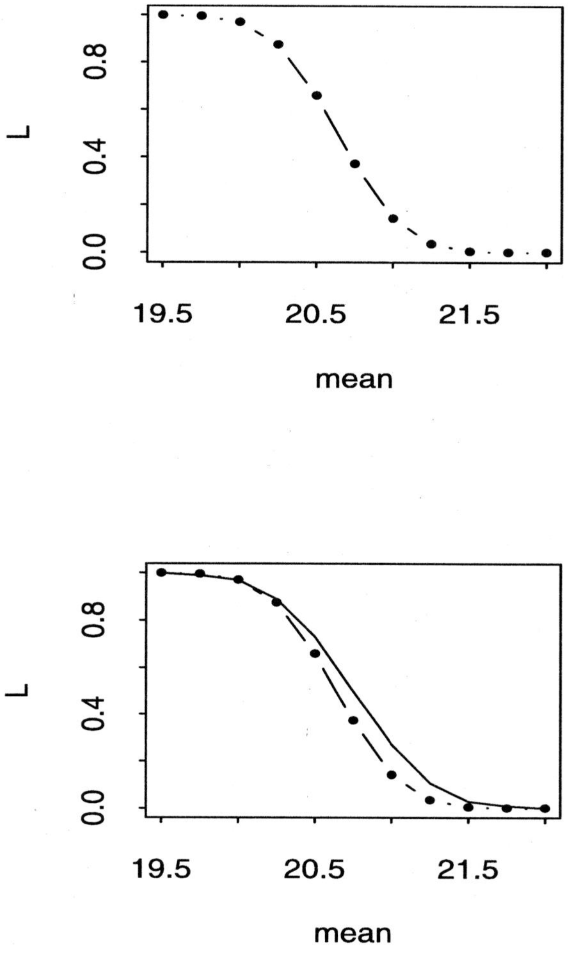

Figure 7.1: (a) OC curve for sample size 50 Exercise 7.85. (b) Both OC Curves.

(a) The sample size is large so the 95 % confidence interval is

$$\bar{x} \pm z_{.025} \cdot \frac{s}{\sqrt{n}} = 1.65 \pm 1.96 \cdot \frac{0.82}{\sqrt{180}} = 1.65 \pm 0.1198$$

or 1.530 to 1.770 hours. We are 95 % confident that the mean time to retain a call is between 1.53 and 1.77 hours.

(b) The population mean μ for all patients is unknown so we never know if it is covered by a particular confidence interval.

(c) Before we sample, the probability is .95 that the interval will cover μ. By the law of large numbers, or long run relative frequency interpretation of probability, if we take many different samples and calculate a 95 % confidence interval for each, about 95% of the time they will cover μ.

7.89 The sample size is $n = 20$ and we would ordinarily report the t-based confidence interval assuming the population is nearly normal. However, the data tell us otherwise. In the middle of the plot, where the observations from a normal sample should be most crowded, there is very large gap. It looks like maybe two populations have been mixed together. If possible, the scientist should trace the history of each specimen to explain this difference. Another possibility is that the scientist made some kind of adjustment to the machine in the middle of the testing. To investigate this possibility, we would need to know the time order of the observations.

Chapter 8

COMPARING TWO TREATMENTS

8.1 1. *Null hypothesis* $H_0 : \mu_1 - \mu_2 = 0$
Alternative hypothesis $H_1 : \mu_1 - \mu_2 < 0$

2. *Level of significance:* $\alpha = 0.05$.

3. *Criterion:* The null hypothesis specifies $\delta = \mu_1 - \mu_0 = 0$. Since the samples are large, we use the large sample statistic where we estimate each population variance by the sample variance.

$$Z = \frac{\overline{X} - \overline{Y} - \delta}{\sqrt{\frac{S_1^2}{n_1} + \frac{S_2^2}{n_2}}}$$

The alternative is one-sided so we reject the null hypothesis for $Z < -z_{.05} = -1.645$

4. *Calculations:* Since $n_1 = 30$, $n_2 = 40$, $\overline{x} = 1908.8$, $\overline{y} = 2114.3$, $s_1 = 327.1$, and $s_2 = 472.3$

$$\sqrt{\frac{327.1^2}{30} + \frac{472.3^2}{40}} = 95.62$$

and

$$z = \frac{1908.8 - 2114.3}{95.62} = -2.15 < -1.645,$$

5. *Decision:* Because $-2.15 < -1.645$, we reject the null hypothesis at the .05 level of significance. The P-value $.0158 = P[Z < -2.15]$ gives strong support for rejecting the null hypothesis.

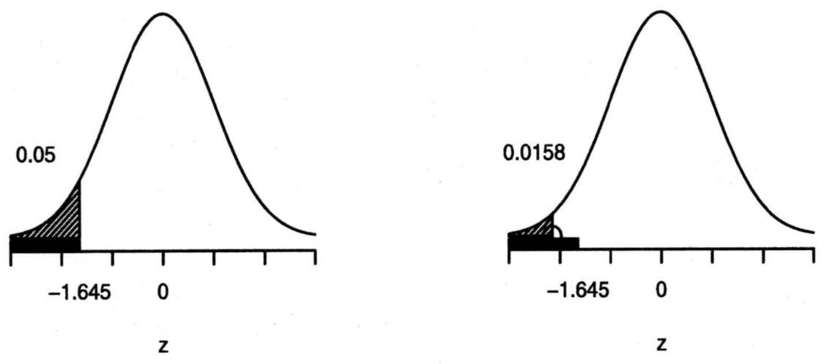

(a) Rejection region (b) P-value for Exercise 8.1

8.3 1. *Null hypothesis* $H_0 : \mu_1 - \mu_2 = 0$

Alternative hypothesis $H_1 : \mu_1 - \mu_2 \neq 0$

2. *Level of significance:* $\alpha = 0.05$.

3. *Criterion:* The null hypothesis specifies $\delta = \mu_1 - \mu_0 = 0$. Since the samples are large, we use the large sample statistic where we estimate each population variance by the sample variance.

$$Z = \frac{\overline{X} - \overline{Y} - \delta}{\sqrt{\frac{S_1^2}{n_1} + \frac{S_2^2}{n_2}}}$$

The alternative is two-sided so we reject the null hypothesis for $Z > z_{.025}$ or $Z < -z_{.025}$

4. *Calculations:* Since $n_1 = 33$, $n_2 = 31$, $\overline{x} = 115.1$, $\overline{y} = 114.6$, $s_1 = 0.47$, and $s_2 = 0.38$

$$\sqrt{\frac{.47^2}{33} + \frac{0.38^2}{31}} = 0.10655$$

and

$$z = \frac{115.1 - 114.6}{0.10655} = 4.69 > 1.96,$$

5. *Decision:* Because $4.69 > 1.96$, we reject the null hypothesis at the .05 level of significance. The P-value $2P[Z > 4.69]$ rounds to 0.00000 and gives extremely strong support for rejecting the null hypothesis.

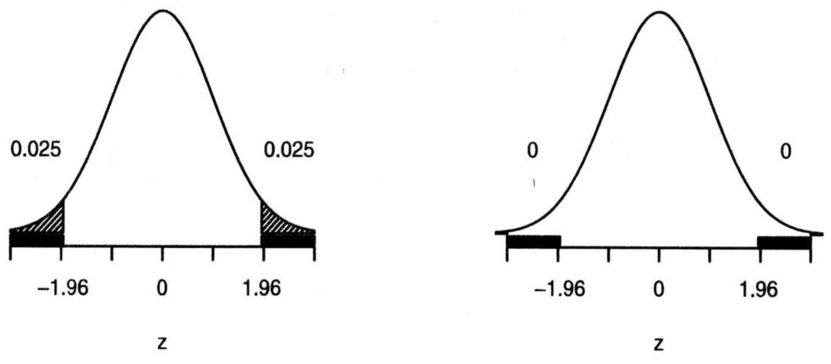

(a) Rejection region (b) P-value for Exercise 8.3

8.5 (a) 1. *Null hypothesis* $H_0 : \mu_1 - \mu_2 = 0$

Alternative hypothesis $H_1 : \mu_1 - \mu_2 \neq 0$

2. *Level of significance:* $\alpha = 0.05$.

3. *Criterion:* The null hypothesis specifies $\delta = \mu_1 - \mu_0 = 0$. Since the samples are large, we use the large sample statistic where we estimate each population variance by the sample variance.

$$Z = \frac{\overline{X} - \overline{Y} - \delta}{\sqrt{\frac{S_1^2}{n_1} + \frac{S_2^2}{n_2}}}$$

The alternative is two-sided so we reject the null hypothesis for $Z > z_{.025}$ or $Z < -z_{.025}$

4. *Calculations:* Since $n_1 = 75$, $n_2 = 75$, $\overline{x} = 83.2$, $\overline{y} = 90.8$, $s_1 = 19.3$, and $s_2 = 21.4$

$$\sqrt{\frac{19.3^2}{75} + \frac{21.4^2}{75}} = 3.3276$$

and

$$z = \frac{83.2 - 90.8}{3.3276} = -2.28 < -1.96,$$

5. *Decision:* Because $-2.28 < -1.96$, we reject the null hypothesis at the .05 level of significance. The P-value $.0226 = 2P[Z < -2.28]$ gives strong support for rejecting the null hypothesis.

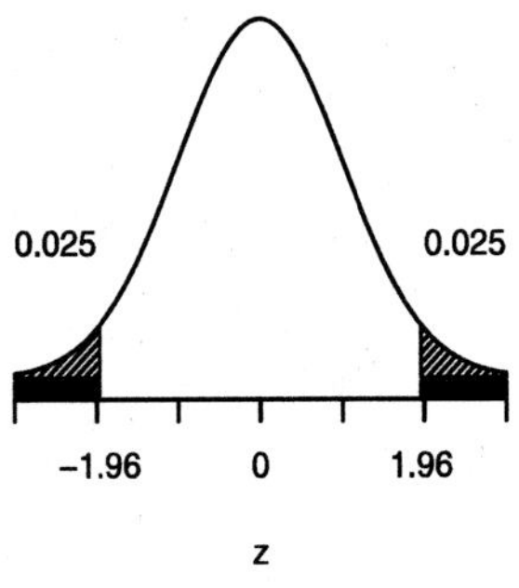

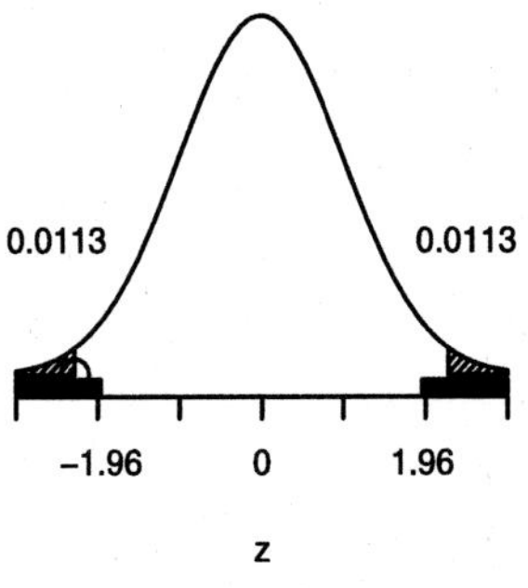

(a) Rejection region (b) P-value for Exercise 8.5

(b) We estimate the fictitious single variance σ^2 and sample size n by

$$\sigma^2 = s_1^2 + s_2^2 = 19.3^2 + 21.4^2 = 830.45 \qquad n = \frac{s_1^2 + s_2^2}{\frac{s_1^2}{n_1} + \frac{s_2^2}{n_2}} = \frac{19.3^2 + 21.4^2}{\frac{19.3^2}{75} + \frac{21.4^2}{75}} = 75$$

Since the alternative is two-sided, and $\delta_0 = 0$, we calculate

$$-z_{\alpha/2} + \sqrt{n}\,\frac{(\delta_0 - \delta')}{\sigma} = -1.96 + \sqrt{75}\,\frac{(0-(-12))}{\sqrt{830.45}} = 1.646$$

$$z_{\alpha/2} + \sqrt{n}\,\frac{(\delta_0 - \delta')}{\sigma} = 1.96 + \sqrt{75}\,\frac{(0-(-12))}{\sqrt{830.45}} = 5.566$$

The power at $\delta" = 12$ is $P(Z < 1.646) + P(Z > 5.566) = 0.950$ so the probability of failing to reject H_0 is $\beta = 1 - 0.950 = 0.050$

8.7 (a) We have $\overline{x} = (6+2+7)/3 = 5$ and $\overline{y} = (14+10)/2 = 12$ so the deviations are

Deviations x	1	-3	2
Deviations y	2	-2	

(b) The pooled variance is

$$s_p^2 = \frac{\sum_{i=1}^{n_1}(x_i - \overline{x})^2 + \sum_{i=1}^{n_2}(y_i - \overline{y})^2}{n_1 + n_2 - 2} = \frac{1^2 + (-3)^2 + 2^2 + 2^2 + (-2)^2}{3+2-2} = \frac{22}{3} = 7.333$$

8.9 1. *Null hypothesis* $H_0 : \mu_1 - \mu_2 = 1.5$
Alternative hypothesis $H_1 : \mu_1 - \mu_2 > 1.5$

2. *Level of significance:* $\alpha = 0.05$.

3. *Criterion:* The null hypothesis specifies $\delta_0 = \mu_1 - \mu_0 = 0$. Since the samples are small, but we can

assume that the populations are normal with the same variance, we use the two-sample t statistic

$$t = \frac{(\bar{X} - \bar{Y}) - \delta_0}{\sqrt{(n_1 - 1)S_1^2 + (n_2 - 1)S_2^2}} \sqrt{\frac{n_1 n_2 (n_1 + n_2 - 2)}{n_1 + n_2}}$$

Since the alternative hypothesis is one-sided, $\delta > 1.5$, we reject the null hypothesis when $t > t_{.05}$ where, from Table 4, with $n_1 + n_2 - 2 = 16$ degrees of freedom, $t_{.05} = 1.746$.

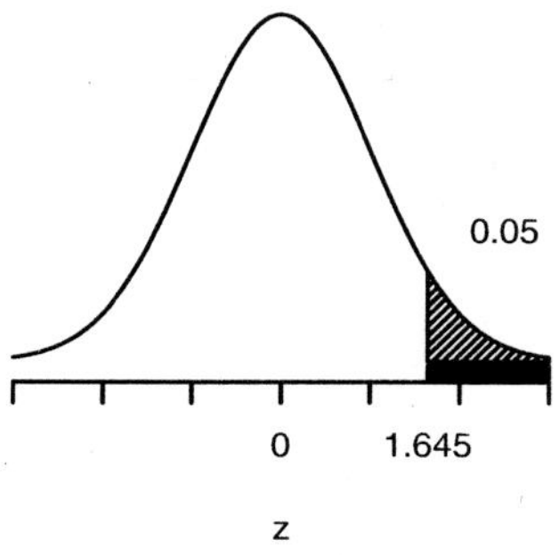

4. *Calculations:* Since $n_1 = 8$, $n_2 = 10$, $\overline{x} = 9.67$, $\overline{y} = 7.43$, $s_1 = 1.81$, and $s_2 = 1.48$

$$t = \frac{(9.67 - 7.43) - 1.5}{\sqrt{7(1.81)^2 + 9(1.48)^2}} \sqrt{\frac{8 \cdot 10 \cdot 16}{18}} = .96,$$

5. *Decision:* We cannot reject the null hypothesis.

8.11 1. Let μ_1 be the mean for California and μ_2 the mean for Oregon.

Null hypothesis $H_0 : \mu_1 - \mu_2 = 0$

Alternative hypothesis $H_1 : \mu_1 - \mu_2 \neq 0$

2. *Level of significance:* $\alpha = 0.01$.

3. *Criterion:* The null hypothesis specifies $\delta_0 = \mu_1 - \mu_0 = 0$. Since the samples are small, but we can assume that the populations are normal with the same variance, we use the two-sample t statistic

$$t = \frac{(\bar{X} - \bar{Y}) - \delta_0}{\sqrt{(n_1 - 1)S_1^2 + (n_2 - 1)S_2^2}} \sqrt{\frac{n_1 n_2 (n_1 + n_2 - 2)}{n_1 + n_2}}$$

Since the alternative hypothesis is two-sided, we reject the null hypothesis when $t < -t_{.005}$ or $t > t_{.005}$ where $t_{.005} = 3.012$ for 13 degrees of freedom.

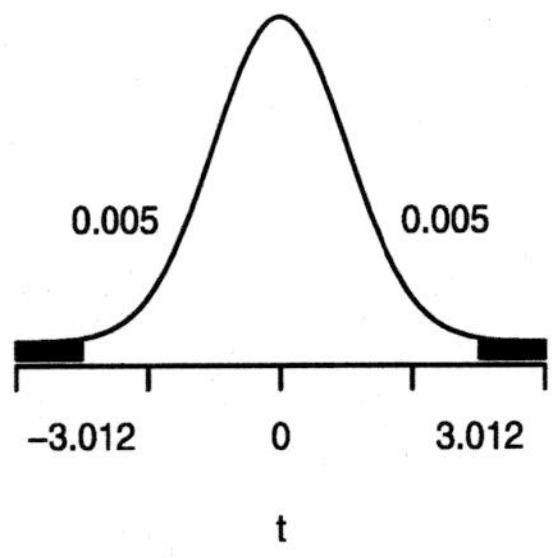

4. *Calculations:* Here $n_1 = 9$ and $n_2 = 6$, and we first calculate $\overline{x} = 58$, $s_1 = 109$, $\overline{y} = 51.833$, $s_2 = 160.97$. Then

$$t = \frac{58 - 51.833}{\sqrt{8(109) + 5(160.97)}}\sqrt{\frac{9 \cdot 6 \cdot 13}{15}} = 1.03.$$

5. *Decision:* We cannot reject the null hypothesis at level of significance $\alpha = .01$.

8.13 (a) 1. Let μ_1 be the mean for Alloy 1 and μ_2 the mean for Alloy 2. first sample has sample variance $s_1^2 = 3.193$ and the second sample has $s_2^2 = 12.135$. While not quite four times as large they are quite disparate so we do not pool.

Null hypothesis $H_0 : \mu_1 - \mu_2 = 0$

Alternative hypothesis $H_1 : \mu_1 - \mu_2 < 0$

2. *Level of significance:* $\alpha = 0.05$.

3. *Criterion:* The null hypothesis specifies $\delta_0 = \mu_1 - \mu_0 = 0$. Since the samples are small, but we cannot assume equal variances, we assume normality and use the approximate two-sample t' statistic

$$t' = \frac{\bar{X} - \bar{Y} - \delta_0}{\sqrt{\frac{S_1^2}{n_1} + \frac{S_2^2}{n_2}}}$$

where the degrees of freedom must be estimated. Since the alternative hypothesis is one-sided, we reject the null hypothesis when $t' < -t_{.05}$

4. *Calculations:* The mean and variance of the first sample are $\overline{x} = 64.67$ and $s_1^2 = 3.193$. The mean and variance of the second sample are $\overline{y} = 66.28$ and $s_2^2 = 12.135$. The degrees of freedom of the test are given by

$$\begin{aligned}\text{df} &= \frac{\left(s_1^2/n_1 + s_2^2/n_2\right)^2}{(s_1^2/n_1)^2/(n_1 - 1) + (s_2^2/n_2)^2/(n_2 - 1)} \\ &= \frac{(3.193/10 + 12.135/10)^2}{(3.193/10)^2/9 + (12.135/10)^2/9} = 13.63 \approx 14\end{aligned}$$

Thus, the criterion is to reject the null hypothesis when $t < t_{.05} = -1.761$.

In this case,

$$t' = \frac{\bar{X} - \bar{Y}}{\sqrt{S_1^2/n_1 + S_2^2/n_2}} = -1.30.$$

5. *Decision* The null hypothesis cannot be rejected at level .05.

(b) The first sample has sample variance $s_1^2 = 597.867$. and second sample has $s_2^2 = 202.000$. Although the first is only three times larger, they are quite disparate and we do not pool.

1. *Null hypothesis* $H_0 : \mu_1 - \mu_2 = 0$

   Alternative hypothesis $H_1 : \mu_1 - \mu_2 \neq 0$

2. *Level of significance:* $\alpha = 0.01$.

3. *Criterion:* The null hypothesis specifies $\delta_0 = \mu_1 - \mu_0 = 0$. Since the samples are small, but we cannot assume equal variances, so we assume normality and use the approximate two-sample t' statistic

$$t' = \frac{\bar{X} - \bar{Y} - \delta_0}{\sqrt{\frac{S_1^2}{n_1} + \frac{S_2^2}{n_2}}}$$

   where the degrees of freedom must be estimated. Since the alternative hypothesis is two-sided, we reject the null hypothesis when $t' < -t_{.01}$.

4. *Calculations:* The mean and variance of the first sample are $\overline{x} = 127.33$ and $s_1^2 = 597.867$. The mean and variance of the second sample are $\overline{y} = 129.00$ and $s_2^2 = 202.000$. The degrees of freedom of the test are given by

$$\begin{aligned} \text{df} &= \frac{\left(s_1^2/n_1 + s_2^2/n_2\right)^2}{(s_1^2/n_1)^2/(n_1-1) + (s_2^2/n_2)^2/(n_2-1)} \\ &= \frac{\left(597.867/6 + 202.000/6\right)^2}{(597.867/6)^2/5 + (202.000/6)^2/5} = 8.03 \approx 8 \end{aligned}$$

   Thus, the criterion is to reject the null hypothesis when $|t| > t_{.005} = 3.355$. In this case,

$$t = \frac{\bar{X} - \bar{Y}}{\sqrt{S_1^2/n_1 + S_2^2/n_2}} = -.145.$$

5. *Decision* The null hypothesis cannot be rejected at level .01.

8.15 The sample size is small and we assume the difference has a normal distribution.

1. *Null hypothesis* $H_0 : \mu_D = 0$

   Alternative hypothesis $H_1 : \mu_D \neq 0$

2. *Level of significance:* $\alpha = 0.05$.

3. *Criterion:* We use the paired t statistic

$$t = \frac{\overline{D} - \nu_{D\,0}}{S_D/\sqrt{n}}$$

Since $\alpha = .05$ and the alternative hypothesis is two-sided, we reject the null hypothesis if $t < -t_{.025}$ or if $t > t_{.025}$. There are 4 degrees of freedom so $t_{.025} = 2.776$.

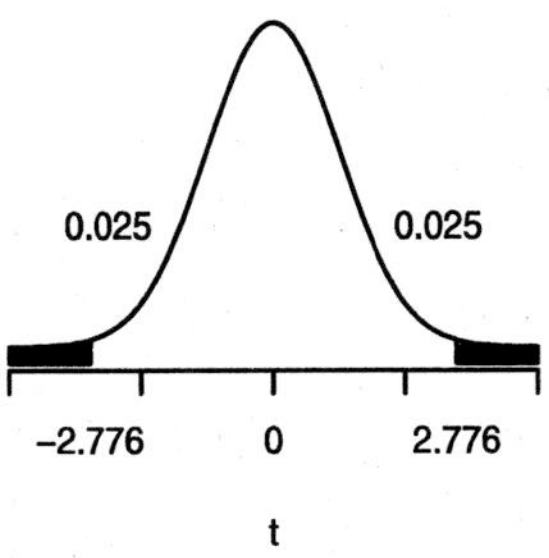

4. *Calculations:* The sample mean of the differences is 1.0 and the variance is 2.0.

$$t = \frac{1.0 - 0}{\sqrt{2.0/5}} = 1.58$$

5. *Decision* We fail to reject the null hypothesis at level of significance .05.

8.17 (a) The sample size is small and we assume the difference has a normal distribution. There are $n = 11$ differences so $t_{.025} = 2.228$ for 10 degrees of freedom. Also, on the square root scale, $\overline{d} = 0.948$ and $s_D = 1.340$. The 95 % confidence interval becomes

$$\overline{d} \pm t_{.025}\,\frac{s_D}{\sqrt{n}} = 0.948 \pm 2.228(\frac{1.340}{\sqrt{11}}) = 0.948 \pm 0.900$$

or $0.048 < \mu_D < 1.848$. We are 95% confident that, on the square root scale, the mean difference in PCB's is between 0.048 and 1.848 $\sqrt{\text{ppb}}$.

(b) 1. *Null hypothesis* $H_0 : \mu_D = 0$

Alternative hypothesis $H_1 : \mu_D \neq 0$

2. *Level of significance:* $\alpha = 0.05$.

3. *Criterion:* The number of pairs is small so we must assume that each difference has a normal distribution. We use the paired t statistic

$$t = \frac{\overline{D} - \mu_{D\,0}}{S_D/\sqrt{n}}$$

Since the alternative hypothesis is two-sided, we reject the null hypothesis if $|t| > t_{.025} = 2.228$.

4. *Calculations:* The difference between The mean of this sample is $\overline{d} = 0.948$ and $s_D = 1.340$. The null distribution specifies $\mu_D = 0$ so

$$t = \frac{0.948 - 0}{1.340/\sqrt{11}} = 2.346$$

5. *Decision:* We reject the null hypothesis at level of significance .05. The P-value is $P[t < -2.346] + P[t > 2.346] = 0.041$.

   Note that the value $\mu_D = 0$ is outside of the 95 % confidence interval in Part (a) so that value should be rejected as a null hypothesis at level $\alpha = 0.05$.

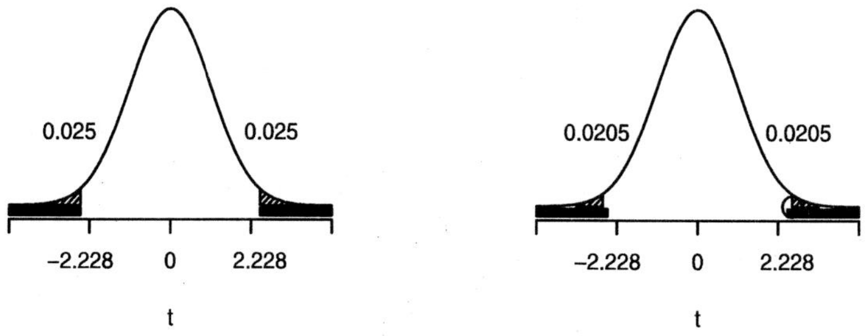

(a) Rejection region (b) P-value for Exercise 8.17

(c) The dot diagram shows that the transformation pulls the two outliers closer to the other observations but they are still a little removed.

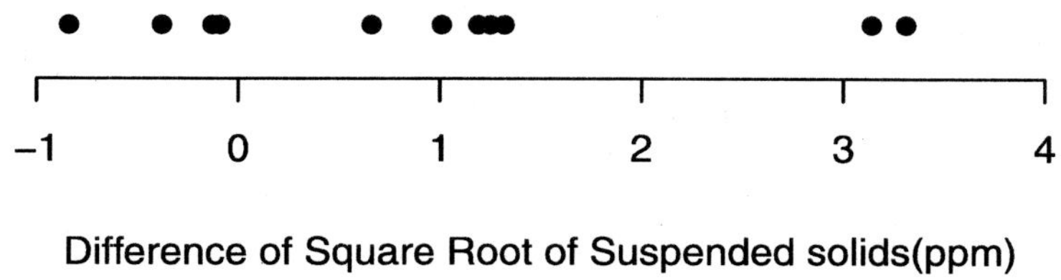

8.19 The sample size $n = 43$ is large and $z_{.025} = 1.96$. We are given $\overline{d} = 0.7$ and $s_D^2 = 3.8$. The large sample 95 % confidence interval becomes

$$\overline{d} \pm z_{.025} \frac{s_D}{\sqrt{n}} = 0.7 \pm 1.96(\frac{\sqrt{3.8}}{\sqrt{43}}) = 0.7 \pm 0.583$$

or $0.117 < \mu_D < 1.283$. We are 95% confident that the mean difference in scores is between 0.12 and 1.28.

8.21 1. *Null hypothesis* $H_0 : \mu_D = 0$

Alternative hypothesis $H_1 : \mu_D > 0$

2. *Level of significance:* $\alpha = 0.01$.

3. *Criterion:* The number of pairs is moderately small so we must assume that each difference has a normal distribution. We use the paired t statistic

$$t = \frac{\bar{D} - \mu_{D\,0}}{S_D/\sqrt{n}}$$

Since $\alpha = .01$ and the alternative hypothesis is one-sided, we reject the null hypothesis if $t > t_{.01}$. There are 15 degrees of freedom so $t > t_{.01} = 2.602$.

4. *Calculations:* The sample mean of the differences is 4.0625 and the variance is 16.996.

$$t = \frac{4.0625 - 0}{\sqrt{16.996/16}} = 3.94$$

5. *Decision* We reject the null hypothesis at level of significance .01. Thus, the physical exercise program is effective. The $P-$value is less than .005 so the evidence against the null hypothesis is strong.

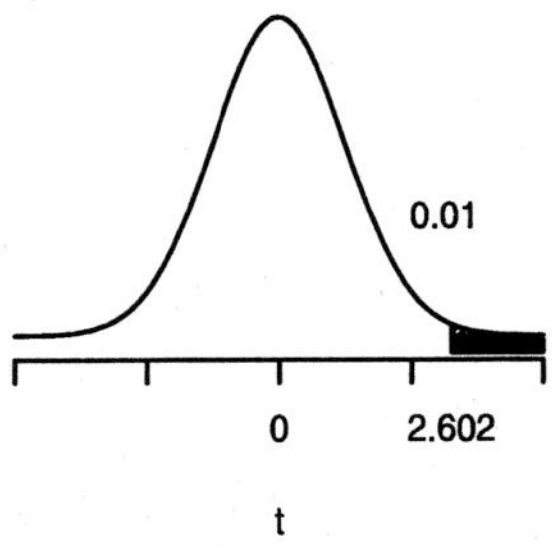

(a) Rejection region

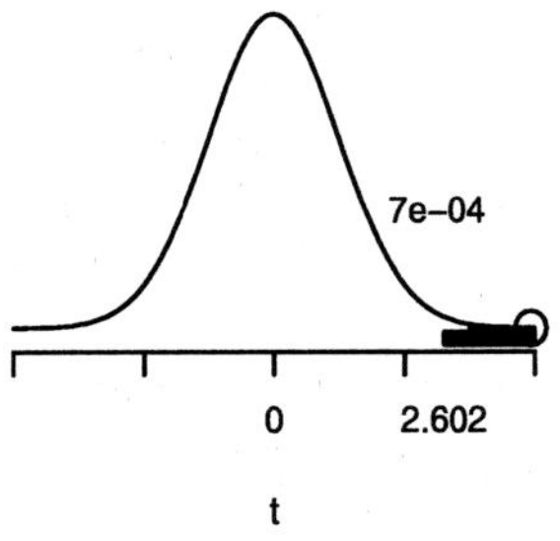

(b) P-value for Exercise 8.21

8.23 (a) First, order the elevators from 1 to 6. Choose a column from Table 7. Starting somewhere in the column and going down the page. Select 3 different numbers by discarding 0 and numbers larger than 6. These are the numbers for the elevators in which the modified circuit board will be inserted.

(b) For each of the six elevators, toss a coin to decide which board to insert first. If heads, the elevator gets the modified circuit board first. After some time, it is replaced by the original board. If tails, elevator gets the modified circuit board second.

8.25 First, order the cars from 1 to 50. Choose two columns from Table 7. Starting somewhere in the column and going down the page. Select 25 different numbers by discarding 00 and numbers larger than 50. These are the numbered cars in which to install the modified air pollution device.

8.27 The 90% small sample confidence interval for $\mu_1 - \mu_2$ is

$$(.42 - .53) \pm 1.782\sqrt{\frac{6(.0072) + 6(.0044)}{12}}\sqrt{\frac{14}{7 \cdot 7}}$$

$$= -.11 \pm .073 = -.183, \quad -.037$$

or $-.183 < \mu_1 - \mu_2 < -.037$. We are 90 % confidence that the difference in means is between -0.183 and -0.037. The second tube has a larger mean charge.

8.29 1. *Null hypothesis* $H_0 : \mu_1 - \mu_2 = 0$

Alternative hypothesis $H_1 : \mu_1 - \mu_2 \neq 0$

2. *Level of significance:* We take $\alpha = 0.05$.

3. *Criterion:* We assume these are samples from two normal populations with the same variance and so we use the two-sample t statistic with $\delta = 0$ as specified by the null hypothesis.

$$t = \frac{(\bar{X} - \bar{Y})}{\sqrt{(n_1 - 1)S_1^2 + (n_2 - 1)S_2^2}}\sqrt{\frac{n_1 n_2(n_1 + n_2 - 2)}{n_1 + n_2}}$$

The alternative is two-sided so we reject the null hypothesis when $|t| > t_{.025}$ with $n_1 + n_2 - 2 = 8$ degrees of freedom. From Table 4, we reject if $|t| > 2.306$.

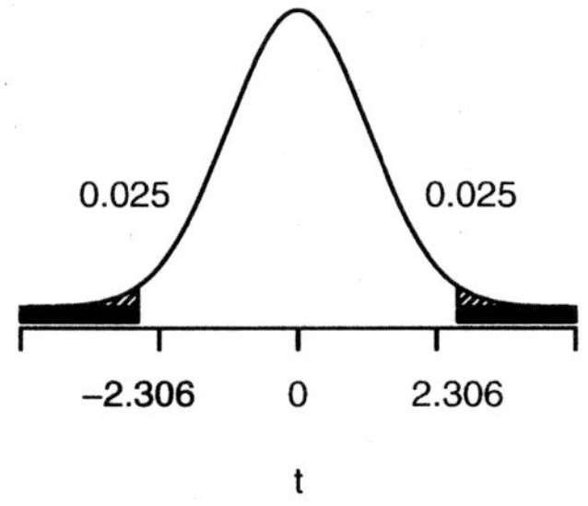

4. *Calculations:* Since $\overline{x} = .33$, $s_1^2 = .0028$, $\overline{y} = .25$, and $s_2^2 = .0032$,

$$t = \frac{(.33 - .25) - 0}{\sqrt{2(.0028) + 6(.0032)}}\sqrt{\frac{3 \cdot 7 \cdot 8}{10}} = 2.082,$$

5. *Decision:* Because $2.082 < 2.306$, we fail to reject the null hypotheses of no difference in mean copper content, at the .05 level of significance.

8.31 The alternative hypothesis is one-sided at the level $\alpha = .01$. We use the computer calculation described in Exercise 7.74 but with

difference $= \delta_0 - \delta = 66.7 - 53.2 = 13.5$ fictitious variance $\sigma^2 = \sigma_1^2 + \sigma_2^2 = 10.8^2 + 14.4^2 = 324$

so $\sigma = 18$. The sample size should be no smaller than 24.

```
Testing mean = null (versus > null)
Calculating power for mean = null + difference
Alpha = 0.01  Assumed standard deviation = 18

                    Sample  Target
    Difference    Size   Power  Actual Power
          13.5      24    0.91      0.911153
```

8.33 (a) Use Table 7 to select 10 cars to install the modified spark plugs. Install the regular plugs to the other cars.

(b) Number the specimens from 1 to 15 and use Table 7 to select 8 numbers between 1 and 15. Use the new oven to bake the specimens having these numbers. Bake the other specimens in the old oven.

8.35 The MINITAB output, for the one sided-test using the data in 7.68, is

```
Two-sample T for Method A vs Method B

           N      Mean     StDev   SE Mean
Method A  10     70.00      3.37       1.1
Method B  10     74.00      5.40       1.7

Difference = mu Method A - mu Method B
Estimate for difference:  -4.00
95% upper bound for difference: -0.51
T-Test of difference = 0 (vs <): T-Value = -1.99
    P-Value = 0.031  DF = 18
Both use Pooled StDev = 4.50
```

Since the P-value (=0.031) is less than .05, we reject the null hypothesis with 95% of confidence. The conclusion is the same as in Exercise 8.10: method B is more effective.

Chapter 9

INFERENCES CONCERNING VARIANCES

9.1 (a) There are $n = 9$ observations sample variance s^2 is given by

$$s^2 = \frac{1}{n-1}\sum(x_i - \bar{x})^2 = \frac{556}{8} = 69.5$$

Thus, the sample standard deviation is $\sqrt{69.5} = 8.34$.

(b) The minimum observation is 102. The maximum observation is 128. thus, the range is 26. Since the sample size is 9, the expected length of the range is 2.970σ. Thus, we estimate σ by $26/2.970 = 8.75$. The percentage difference is

$$\frac{8.75 - 8.34}{8.34} \times 100 = 4.9\,\%$$

9.3 (a) The sample standard deviation is 1.787.

(b) The range of the data is $68.4 - 61.8 = 6.6$. The sample size is 10, so the estimate of the standard deviation is $6.6/3.078 = 2.144$. The relative difference is

$$\frac{2.144 - 1.787}{1.787} \cdot 100 = 20.0 \text{ percent.}$$

9.5 The sample variance is .025. Since the sample size is 9, the $\chi^2_{.01/2}$ with 8 degrees of freedom is 14.860 and $\chi^2_{1-.01/2}$ with 8 degrees of freedom is .207. Thus, if the data are from a normal population, a 99 percent confidence interval for σ^2 is

$$\frac{8(.025)}{14.860} < \sigma^2 < \frac{8(.025)}{.207}$$

or

$$.0067 < \sigma^2 < .483,$$

and the 99 percent confidence interval for the standard deviation is

$$.082 < \sigma < .695$$

9.7 Assuming a normal population, we use the statistic

$$\chi^2 = \frac{(n-1)S^2}{\sigma_0^2}.$$

Since the alternative is $\sigma > 600$, we reject the null hypothesis $\sigma = 600$ if $\chi^2 > \chi^2_{.05}$ with 5 degrees of freedom or if $\chi^2 > 11.076$. In this case, $s = 648$ so the test statistic

$$\chi^2 = \frac{5 \cdot (648)^2}{600^2} = 5.832.$$

Thus, we cannot reject the null hypothesis at the .05 level of significance.

9.9 Since the data are from a normal population, we can use the statistic

$$\chi^2 = \frac{(n-1)S^2}{\sigma_0^2}.$$

The null hypothesis is $\sigma = 15.0$ and the alternative is $\sigma > 15.0$. Since the sample size is 75, we reject the null hypothesis if $\chi^2 > \chi^2_{.05}$ with 74 degrees of freedom Thus, we reject the null hypothesis $\sigma = 15.0$ if $\chi^2 > 95.08$. In this case, $s = 19.3$ minutes and the test statistic

$$\chi^2 = \frac{74 \cdot (19.3)^2}{(15.0)^2} = 122.5$$

so we reject the null hypothesis $\sigma = 15.0$ in favor of the alternative $\sigma > 15.0$, at the .05 level of significance. The $P-$ value is less than 0.005 so the evidence against the null hypothesis is very strong (actual value 0.0003).

9.11 The sample standard deviation is $s = 1.32$ and the sample size is 10. If the data are from a normal population, we can use the statistic

$$\chi^2 = \frac{(n-1)S^2}{\sigma_0^2}.$$

The null hypothesis is $\sigma = 1.20$ and the alternative is $\sigma > 1.20$. Thus, we reject the null hypothesis when $\chi^2 > \chi^2_{.05}$ with 9 degrees of freedom or when $\chi^2 > 16.919$. In this case $s = 1.32$ so

$$\chi^2 = \frac{9 \cdot (1.32)^2}{(1.20)^2} = 10.89$$

Thus, we cannot reject the null hypothesis at the .05 level of significance.

9.13 Since Exercise 8.9 states that the two samples can be assumed to be from normal populations, we can use the statistic

$$F = \frac{S_M^2}{S_m^2}$$

which has an F distribution with $n_M - 1$ and $n_m - 1$ degrees of freedom. The null hypothesis is $\sigma_1^2 = \sigma_2^2$ and the alternative hypothesis is $\sigma_1^2 \neq \sigma_2^2$. The sample sizes are $n_M = 8$ and $n_m = 10$. Thus, we reject the null hypothesis when $F > F_{.02/2}(7, 9)$ or when $F > 5.61$. In this case $s_M^2 = (1.81)^2$ and $s_m^2 = (1.48)^2$ so

$$F = \left(\frac{1.81}{1.48}\right)^2 = 1.496.$$

Thus, we cannot reject the null hypothesis at the .02 level of significance.

9.15 The null hypothesis $\sigma_1^2 = \sigma_2^2$ and the alternative hypothesis is $\sigma_1^2 < \sigma_2^2$. Therefore, we use the statistic

$$F = \frac{S_2^2}{S_1^2}$$

which has an F distribution with $n_2 - 1 = 20$ and $n_1 - 1 = 14$ degrees of freedom. The null hypothesis $\sigma_1^2 = \sigma_2^2$ will be rejected favor of the alternative hypothesis $\sigma_1^2 < \sigma_2^2$ if $F > F_{.01}(20, 14) = 3.51$. In this case

$$F = \left(\frac{4.2}{2.7}\right)^2 = 2.42$$

so we cannot reject the null hypothesis at the .01 level of significance. This analysis assumes that the two samples come from normal populations and that the samples are independent.

9.17 The sample size is 5 and $\chi^2_{.05/2}$ with 4 degrees of freedom is 11.143 and $\chi^2_{1-.05/2}$ with 4 degrees of freedom is .484. The sample standard deviation is 5.7. Thus, the 95 percent confidence interval for σ is

$$\sqrt{\frac{4 \cdot (5.7)^2}{11.143}} < \sigma < \sqrt{\frac{4 \cdot (5.7)^2}{.484}}$$

or

$$3.42 < \sigma < 16.39$$

9.19 The sample size is 101 and $\chi^2_{.05/2}$ with 100 degrees of freedom is 129.561 and $\chi^2_{1-.05/2}$ with 100 degrees of freedom is 74.222. The null hypothesis is $\sigma^2 = .18$ and the alternative is $\sigma^2 \neq .18$. The sample variance is .13. If the data are from a normal population, we can use the statistic

$$\chi^2 = \frac{(n-1)S^2}{\sigma_0^2}.$$

We reject the null hypothesis if $\chi^2 > \chi^2_{.05/2} = 129.561$ or when $\chi^2 < \chi^2_{1-.05/2} = 74.222$. In this case,

$$\chi^2 = \frac{(100)(.13)}{.18} = 72.22.$$

Thus, we reject the null hypothesis $\sigma^2 = .18$, in favor of the alternative $\sigma^2 \neq .18$, at the .05 level of significance. The inspector is not making satisfactory measurements.

9.21 The standard deviation of the first sample is 1.80 and the sample size is 58. The standard deviation of the second sample is 2.42 and the sample size is 27. The null hypothesis $\sigma_1^2 = \sigma_2^2$ and the alternative hypothesis is $\sigma_1^2 \neq \sigma_2^2$. Normal-scores plots of these samples fail to contradict the normal assumption so we use the statistic

$$F = \frac{S_M^2}{S_m^2}$$

which has an F distribution with $n_M - 1 = 26$ and $n_m - 1 = 57$ degrees of freedom. The null hypothesis $\sigma_1^2 = \sigma_2^2$ will be rejected in favor of the alternative hypothesis $\sigma_1^2 \neq \sigma_2^2$ if $F > F_{.01}(26, 57) = 2.11$. In this case

$$F = \left(\frac{2.42}{1.80}\right)^2 = 1.81$$

so we cannot reject the null hypothesis at the .02 level of significance.

9.23 We enter .025 and obtain

```
F distribution with 7 DF in numerator and 4 DF in denominator


 P( X <=x )          x
      0.025   0.181074
```

We see that $1/5.52259 = 0.181074$ checking the result.

Chapter 10

INFERENCES CONCERNING PROPORTIONS

10.1 The large sample 95% confidence interval for p, obtained by substituting $x/n = 84/200 = = .42$ and $z_{\alpha/2} = 1.96$, is

$$.42 - 1.96\sqrt{\frac{(.42)(.58)}{200}} < p < .42 + 1.96\sqrt{\frac{(.42)(.58)}{200}}$$

or $.352 < p < .488$. We are 95 % confident that the true proportion of claims is between .352 and .488.

10.3 The sample proportion is $231/400 = .578$. Using the large sample formula with $x/n = 231/400 = .578$ and $z_{\alpha/2} = 2.575$ gives the 99% confidence interval

$$.578 - 2.575\sqrt{\frac{(.578)(.422)}{400}} < p < .578 + 2.575\sqrt{\frac{(.578)(.422)}{400}}$$

or $.514 < p < .642$. WE are 99 % confident that the proportion of accidents at least partially due to unsafe working conditions is between .514 and .642.

10.5 Using the large sample formula with $x/n = 15/90 = .1667$ and $z_{\alpha/2} = z_{.025} = 1.96$ gives the 95% confidence interval

$$.1667 - 1.96\sqrt{\frac{(.1667)(.8333)}{90}} < p < .1667 + 1.96\sqrt{\frac{(.1667)(.8333)}{90}}$$

or $0.090 < p < 0.244$. WE are 95 % confident the proportion of sections that have serious corrosion is between 0.090 and 0.244.

10.7 Using the large sample error bound gives

$$.065 = z_{\alpha/2}\sqrt{\frac{(.18)(1-.18)}{100}}.$$

Thus, $z_{\alpha/2} = 1.69$. This corresponds to $\alpha = .091$. Therefore, we have 90.9% confidence.

10.9 Using the formula for unknown p with $E = .06$ and $z_{\alpha/2} = z_{.025} == 1.96$ gives

$$n = \frac{1}{4}\left(\frac{1.96}{.06}\right)^2 = 266.78.$$

The required sample size is 267.

10.11 Using the formula for unknown p with $E = .035$ and $z_{\alpha/2} = z_{.005} = 2.575$ gives

$$n = \frac{1}{4}\left(\frac{2.575}{.035}\right)^2 = 1353.2.$$

The required sample size is 1354.

10.13 The exact 95 % confidence interval is

```
 X   N  Sample p          95% CI
16  20  0.800000  (0.563386, 0.942666)
```

10.15 We need to solve the following inequalities for p:

$$-z_{\alpha/2} < \frac{x-np}{\sqrt{np(1-p)}} < z_{\alpha/2}.$$

These inequalities are equivalent to

$$\frac{x^2 - 2npx + n^2p^2}{np(1-p)} < z^2_{\alpha/2},$$

which is equivalent to

$$x^2 - 2npx + n^2p^2 < npz^2_{\alpha/2} - np^2z^2_{\alpha/2}.$$

Combining terms with the like powers of p gives

$$(n^2 + nz^2_{\alpha/2})p^2 - (2nx + nz_{\alpha/2})p + x^2 < 0. \qquad (10.1)$$

Now, changing the inequality to equality in equation (10.1), the two roots of the quadratic equation are

$$p = \frac{(2nx + nz^2_{\alpha/2}) \pm \sqrt{(2nx + nz^2_{\alpha/2})^2 - 4(n^2 + nz^2_{\alpha/2})x^2}}{2(n^2 + nz^2_{\alpha/2})}.$$

Simplifying the terms under the square root gives

$$p = \frac{(2nx + nz^2_{\alpha/2}) \pm \sqrt{4n^2 z^2_{\alpha/2}\left(x(n-x)/n + z^2_{\alpha/2}/4\right)}}{2(n^2 + nz^2_{\alpha/2})}.$$

Thus,

$$p = \frac{x + z^2_{\alpha/2}/2 \pm z_{\alpha/2}\sqrt{x(n-x)/n + z^2_{\alpha/2}/4}}{n + z^2_{\alpha/2}}. \tag{10.2}$$

Let the two roots be p_1 and p_2 respectively with $p_1 < p_2$. Then, equation (10.1) is equivalent to

$$(p - p_1)(p - p_2) < 0.$$

Since $p_1 < p_2$, we have

$$p_1 < p < p_2.$$

Thus the two roots given by equation (10.2) are the upper and lower $100(1-\alpha)\%$ confidence limits for p.

10.17 The sample proportion is $533/4063 = .131$. Using the large sample confidence interval with $z_{\alpha/2} = z_{.025}$ $=1.96$ gives

$$.131 - 1.96\sqrt{\frac{(.131)(.869)}{4063}} < p < .131 + 1.96\sqrt{\frac{(.131)(.869)}{4063}}$$

or $.12 < p < .14$ as the 95% confidence interval. We are 95 % confident that the proportion of defective gray paint jobs is between 0.12 and 0.14.

10.19
1. *Null hypothesis* $H_0 : p = .3$
   Alternative hypothesis $H_1 : p > .3$

2. *Level of significance:* $\alpha = 0.05$.

3. *Criterion:* Using a normal approximation for the binomial distribution, we reject the null hypothesis when

$$Z = \frac{X - np_0}{\sqrt{np_0(1-p_0)}} > z_{.05}.$$

Since $\alpha = .05$ and $z_{.05} = 1.645$, the null hypothesis must be rejected if $Z > 1.645$.

4. *Calculations:* $p_0 = .3,\ X = 47$, and $n = 120$ so

$$Z = \frac{47 - 120(.30)}{\sqrt{120(.30)(.70)}} = 2.19.$$

5. *Decision:* Since the observed value $2.19 > z_{.05} = 1.645$, we reject the null hypothesis at the 5% level of significance. The evidence against the null hypotheses is quite strong since the P-value is .0143.

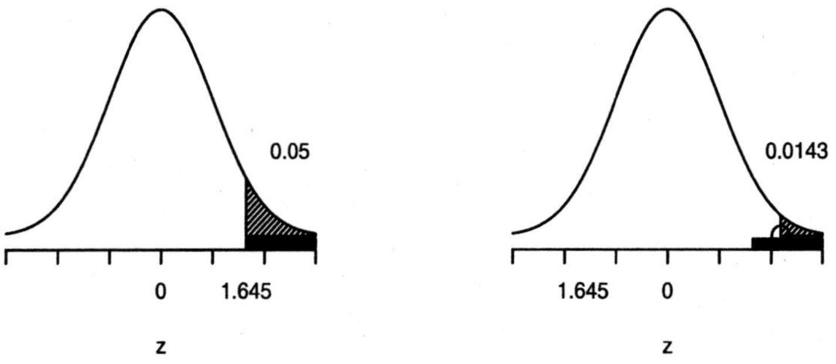

(a) Rejection Region (b) P-value for Problem 9.19

10.21 1. *Null hypothesis* $H_0 : p = .4$

Alternative hypothesis $H_1 : p < .4$

2. *Level of significance:* $\alpha = 0.01$.

3. *Criterion:* Using a normal approximation for the binomial distribution, we reject the null hypothesis when

$$Z = \frac{X - np_0}{\sqrt{np_0(1 - p_0)}} < -z_{.01}.$$

Since $\alpha = .01$ and $z_{.01} = 2.33$, the null hypothesis must be rejected if $Z < -2.33$.

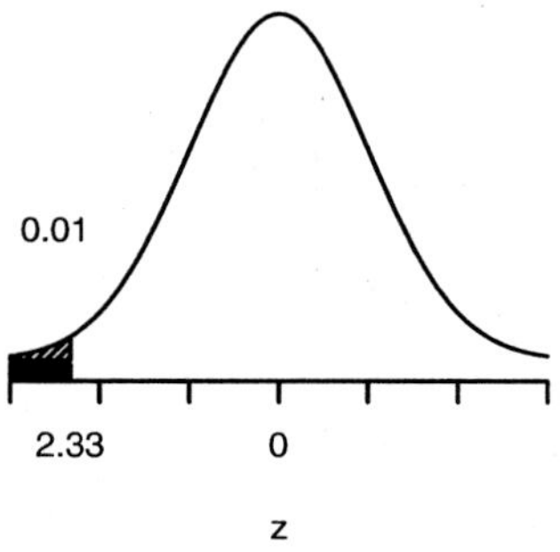

4. *Calculations:* $p_0 = .4$, $X = 49$, and $n = 150$ so

$$Z = \frac{49 - 150(.40)}{\sqrt{150(.40)(.60)}} = -1.83.$$

5. *Decision:* Since the observed value $-1.83 > -z_{.01} = -2.33$, we cannot reject the null hypothesis at the 1% level of significance.

10.23 1. *Null hypothesis* $H_0 : p = .06$

Alternative hypothesis $H_1 : p > .06$

2. *Level of significance:* $\alpha = 0.05$.

3. *Criterion:* Using a normal approximation for the binomial distribution, we reject the null hypothesis when

$$Z = \frac{X - np_0}{\sqrt{np_0(1-p_0)}} > z_{.05}.$$

Since $\alpha = .05$ and $z_{.05} = 1.645$, the null hypothesis must be rejected if $Z > 1.645$.

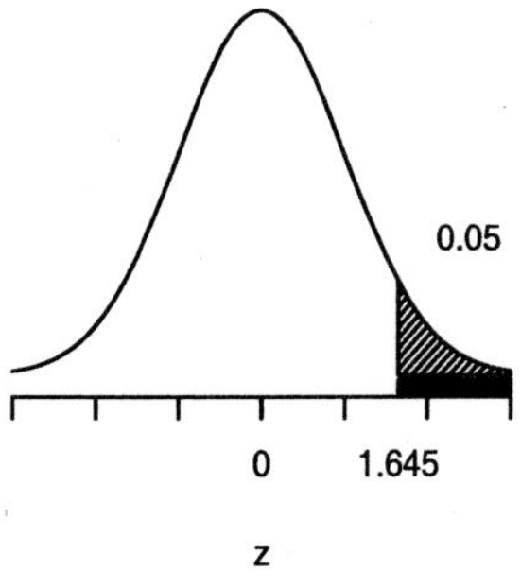

4. *Calculations:* $p_0 = .06$, $X = 17$, and $n = 200$ so

$$Z = \frac{17 - 200(.06)}{\sqrt{200(.06)(.94)}} = 1.489.$$

5. *Decision:* Since the observed value $1.489 < z_{.05} = 1.645$, we cannot reject the null hypothesis at the 5% level.

10.25 1. The alternative is what we intend to establish. *Null hypothesis* $H_0 : p = .95$

Alternative hypothesis $H_1 : p < .95$

2. *Level of significance:* $\alpha = 0.05$.

3. *Criterion:* Using a normal approximation for the binomial distribution, we reject the null hypoth-

esis when

$$Z = \frac{X - np_0}{\sqrt{np_0(1-p_0)}} \quad < \quad -z_{.05}.$$

Since $\alpha = .05$ and $z_{.05} = 1.645$, the null hypothesis must be rejected if $Z < -1.645$.

4. *Calculations:* $p_0 = .95$, $X = 937$, and $n = 1000$ so

$$Z = \frac{937 - 1000(.95)}{\sqrt{1000(.95)(.05)}} = -1.886.$$

5. *Decision:* Since the observed value $-1.886 < -z_{.05} = -1.645$, we reject the null hypothesis at the 5% level.

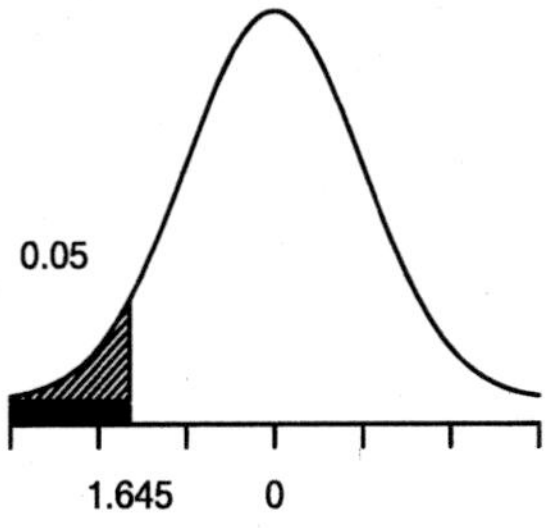

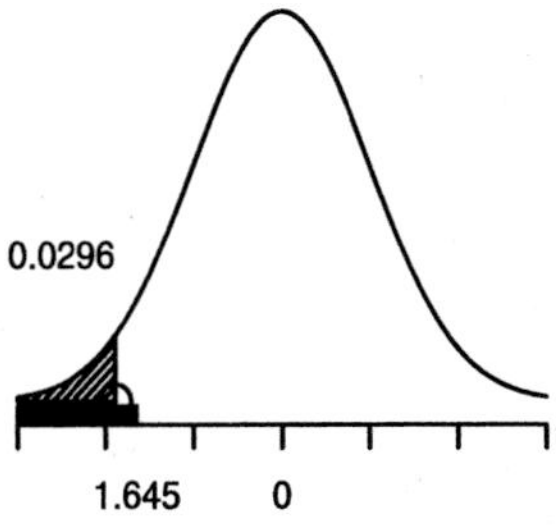

10.27 We use the χ^2 statistic with 4 degrees of freedom to test the null hypothesis that $p_1 = p_2 = p_3 = p_4 = p_5$ against the alternative that at least one of the probabilities is not equal. Thus, we reject the null hypothesis at the 1% level when $\chi^2 > \chi^2_{.01} = 13.277$. In Table 10.1, the expected frequency and the contribution to the χ^2 statistic of each cell are given in the parentheses and brackets respectively.

Table 10.1. Exercise 10.27.

	I	II	III	IV	V	Total
Fail	14 (14.67) [.03]	33 (29.33) [.46]	21 (26.40) [1.10]	17 (17.60) [.02]	25 (22.00) [.41]	110
Pass	86 (85.33) [.01]	167 (170.67) [.08]	159 (153.60) [.19]	103 (102.40) [.00]	125 (128.00) [.07]	640
Total	100	200	180	120	150	750

The χ^2 statistic is

$$\chi^2 = \frac{.67^2}{14.67} + \frac{3.67^2}{29.33} + \frac{5.40^2}{26.40} + \frac{.60^2}{17.60} + \frac{3.00^2}{22.00} +$$

$$\frac{.67^2}{85.33} + \frac{3.67^2}{170.67} + \frac{5.40^2}{153.60} + \frac{.60^2}{102.40} + \frac{3.00^2}{128.00} = 2.37.$$

Thus, we cannot reject the null hypothesis.

10.29 We use the χ^2 statistic with 2 degrees of freedom to test the null hypothesis that the actual proportions are the same against the alternative that they are not the same. Thus, we reject the null hypothesis at the 1% level when $\chi^2 > \chi^2_{.01} = 9.210$. In Table 10.3, the expected frequency and the contribution to the χ^2 statistic of each cell are given in the parentheses and brackets respectively.

Table 10.3. Exercise 10.29.

	Agency I	Agency II	Agency III	Total
For the pension plan	67 (65.00) [.06]	84 (97.50) [1.87]	109 (97.50) [1.36]	260
Against the pension plan	33 (35.00) [.11]	66 (52.50) [3.47]	41 (52.50) [2.52]	140
Total	100	150	150	400

The χ^2 statistic is

$$\begin{aligned}\chi^2 &= \frac{2.00^2}{65.00} + \frac{13.50^2}{97.50} + \frac{11.50^2}{97.50} + \\ &\quad \frac{2.00^2}{35.00} + \frac{13.50^2}{52.50} + \frac{11.50^2}{52.50} = 9.39.\end{aligned}$$

Thus, we reject the null hypothesis.

10.31 With reference to part (b) of the preceding exercise,

$$Z^2 = \frac{(13/250 - 7/250)^2}{(.04)(.96)(2/250)} = 1.875.$$

This verifies the fact that $Z^2 = \chi^2$ in this case.

10.33 Let p_1 and p_2 be proportions of reworking units before and after the training respectively. The 99% confidence interval for the true difference of the proportions, $p_1 - p_2$, is

$$\begin{aligned}&x_1/n_1 - x_2/n_2 \pm z_{\alpha/2}\sqrt{\frac{(x_1/n_1)(1-x_1/n_1)}{n_1} + \frac{(x_2/n_2)(1-x_2/n_2)}{n_2}} \\ &= 26/200 - 12/200 \pm 2.575\sqrt{\frac{(26/200)(1-26/200)}{200} + \frac{(12/200)(1-12/200)}{200}} \\ &= .07 \pm .075\end{aligned}$$

or $-.005 < p_1 - p_2 < .145$. We are 99 % confident that the proportion of units requiring reworking under the new method could be .145 lower to .005 higher than for the old method.

10.35 The 99% confidence interval for the true difference of the proportions is

$$\begin{aligned}
x_1/n_1 - x_2/n_2 \pm z_{\alpha/2} & \sqrt{\frac{(x_1/n_1)(1-x_1/n_1)}{n_1} + \frac{(x_2/n_2)(1-x_2/n_2)}{n_2}} \\
&= 205/250 - 137/250 \\
&\pm 2.575 \sqrt{\frac{(205/250)(1-205/250)}{250} + \frac{(137/250)(1-137/250)}{250}} \\
&= .272 \pm .102
\end{aligned}$$

or $.170 < p_1 - p_2 < .374$. We are 99 % confident that the proportion of successful answers is, from .170 to .374, higher for well prepared students.

10.37 Notice that $n = \sum_j n_j$ and

$$e_{1j} = n_j \frac{x}{n}, \quad e_{2j} = n_j \frac{n-x}{n}.$$

The sum of the expected frequencies of the first row is

$$\sum_j e_{1j} = \sum_j n_j \frac{x}{n} = \frac{x}{n} \sum_j n_j = x.$$

Similarly, the sum of the expected frequencies of the second row is

$$\sum_j e_{2j} = \sum_j n_j \frac{n-x}{n} = \frac{n-x}{n} \sum_j n_j = n - x.$$

Also, the sum of the expected frequencies of the jth column is

$$e_{1j} + e_{2j} = n_j \frac{x}{n} + n_j \frac{n-x}{n} = n_j.$$

10.39 The null hypothesis is that there is homogeneity among the shops' repair distributions. We use the χ^2 statistic with 4 degrees of freedom to test at the 5% level. Thus, we reject the null hypothesis if $\chi^2 > \chi^2_{.05} = 5.991$. In Table 10.5, we write the Shops as rows and give the expected frequency and the contribution to the χ^2 statistic of each cell are given in the parentheses and brackets respectively.

Table 10.5. Exercise 10.39.

	Complete	Adjustment	Incomplete	Total
Shop 1	78 (62.67) [3.752]	15 (25.33) [4.215]	7 (12.00) [2.083]	100
Shop 2	56 (62.67) [0.709]	30 (25.33) [0.860]	14 (12.00) [0.333]	100
Shop 3	54 (62.67) [1.199]	31 (25.33) [1.268]	15 (12.00) [0.750]	100
Total	188	76	36	300

The χ^2 statistic is

$$\begin{aligned}\chi^2 &= \frac{15.33^2}{62.67} + \frac{(-10.33)^2}{25.33} + \frac{(-5)^2}{12} \\ &\quad \frac{(-6.67)^2}{62.67} + \frac{(4.67)^2}{25.33} + \frac{2^2}{12} \\ &\quad \frac{(-8.67)^2}{62.67} + \frac{5.67^2}{25.33}\frac{3^2}{12} = 15.168.\end{aligned}$$

Thus, we reject the null hypothesis of homogeneity among the shops' repair distributions at the 5 % level. The largest contributions to the χ^2 statistic come from the first row where Shop 1 has higher than expected complete repairs and lower than expected adjustments and incomplete repairs.

10.41 To test the null hypothesis that there is no dependence between fidelity and selectivity against the alternative that they is dependence at the 1% level, we use the χ^2 statistic with 4 degrees of freedom and reject the null hypothesis when $\chi^2 > \chi^2_{.01} = 13.277$. In Table 10.7, the expected frequency and the contribution to the χ^2 statistic of each cell are given in the parentheses and brackets respectively.

Table 10.7. Exercise 10.41.

		Fidelity			
		Low	Average	High	Total
Selectivity	Low	6 (13.68) [4.31]	12 (23.16) [5.38]	32 (13.16) [26.98]	50
	Average	33 (30.65) [.18]	61 (51.87) [1.61]	18 (29.47) [4.47]	112
	High	13 (7.66) [3.72]	15 (12.97) [.32]	0 (7.37) [7.37]	28
	Total	52	88	50	190

The χ^2 statistic is

$$\begin{aligned}\chi^2 &= \frac{7.68^2}{13.68} + \frac{11.16^2}{23.16} + \frac{18.84^2}{13.16} + \\ &\quad \frac{2.35^2}{30.65} + \frac{9.13^2}{51.87} + \frac{11.47^2}{29.47} + \\ &\quad \frac{5.34^2}{7.66} + \frac{2.03^2}{12.97} + \frac{7.37^2}{7.37} \\ &= 54.328.\end{aligned}$$

Thus, we reject the null hypothesis and conclude that there is dependence between fidelity and selectivity. The major contributions to χ^2 come from the High Fidelity category. The Low Selectivity and High Fidelity count is very high.

10.43 The mean of the observed distribution is

$$\bar{x} = \frac{0 \cdot 101 + 1 \cdot 79 + 3 \cdot 1}{200} = .6.$$

Thus $.6/4 = .15$ or 15% of the tractors require adjustment. The binomial probabilities and expected frequencies are:

Table 10.9. Exercise 10.43.

No. needing adjustments	Binomial prob. p=.15, n=4	Expected numbers	
0	.5220	104.4	
1	.3685	73.7	
2	.0975	19.5	
3	.0115	2.3	21.9
4	.0005	.1	

The null hypothesis is that the data are from a binomial distribution with $p = .15$ and the alternative is that the data are not from a binomial distribution with $p = .15$. The χ^2 statistic now has 1 degree of freedom, so we reject the null hypothesis at the 1% level when $\chi^2 > \chi^2_{.01} = 6.635$. Now,

$$\chi^2 = \frac{(101-104.4)^2}{104.4} + \frac{(79-73.7)^2}{73.7} + \frac{(20-21.9)^2}{21.9} \quad = \quad .657.$$

Thus, we cannot reject the null hypothesis.

10.45 The distribution function of the exponential distribution with $\mu = 40$ is $F(t) = 1 - \exp(-t/40)$. Thus, we can summarize the data and the expected frequencies in the table.

Table 10.13. Exercise 10.45.

X = Service life	Frequency	Exponential probability	Expected frequency
$X < 20$	46	.393	39.3
$20 \leq X < 40$	19	.239	23.9
$40 \leq X < 60$	17	.145	14.5
$60 \leq X < 80$	12	.088	8.8
$X \geq 80$	6	.135	13.5

We use the χ^2 statistic with $5 - 1 = 4$ degrees of freedom to test the null hypothesis that the data are from an exponential distribution with $\mu = 40$ against the alternative that they are from some other distribution. Thus, we reject the null hypothesis at the 1% level when $\chi^2 > \chi^2_{.01} = 13.227$. Now,

$$\begin{aligned}\chi^2 &= \frac{(46-39.3)^2}{39.3} + \frac{(19-23.9)^2}{23.9} + \frac{(17-14.5)^2}{14.5} + \\ &\quad \frac{(12-8.8)^2}{8.8} + \frac{(6-13.5)^2}{13.5} \quad = \quad 7.91.\end{aligned}$$

Since $7.91 < \chi^2_{.01} = 13.227$, we cannot reject the null hypothesis.

10.47 The MINITAB output is:

```
Expected counts are printed below observed counts

           Above  Average    Below    Total
     1        21       64       17      102
            20.3     63.5     18.2

     2        16       49       14       79
  15.8      49.2     14.1

     3        29       93       28      150
            29.9     93.4     26.7

Total         66      206       59      331

ChiSq =    0.02 +   0.00 +   0.08 +
           0.00 +   0.00 +   0.00 +
   0.03 +   0.00 +   0.06 = 0.20
df = 4
```

We get the same conclusion as in Exercise 9.40.

10.49 1. *Null hypothesis* $H_0 : p = .2$

Alternative hypothesis $H_1 : p < .2$

2. *Level of significance:* $\alpha = 0.05$.

3. *Criterion:* Using a normal approximation for the binomial distribution, we reject the null hypothesis when

$$Z = \frac{X - np_0}{\sqrt{np_0(1-p_0)}} \quad < \quad -z_{.05}.$$

Since $\alpha = .05$ and $z_{.05} = 1.645$, the null hypothesis must be rejected if $Z < -1.645$.

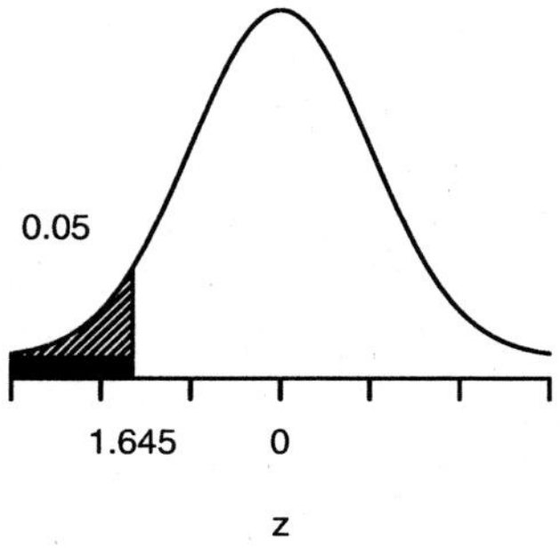

4. *Calculations:* $p_0 = .2$ $X = 18$, and $n = 100$ so

$$Z = \frac{18 - 100(.20)}{\sqrt{100(.20)(.80)}} = -.5.$$

5. *Decision:* Since the observed value $-.5 > -z_{.05} = -1.645$, we cannot reject the null hypothesis at the 5% level of significance.

10.51 1. *Null hypothesis* $H_0 : p = 0.18$
Alternative hypothesis $H_1 : p \neq 0.18$

2. *Level of significance:* $\alpha = 0.01$.

3. *Criterion:* Using a normal approximation for the binomial distribution, we reject the null hypothesis when

$$Z = \frac{X - np_0}{\sqrt{np_0(1-p_0)}} \quad < \quad -z_{.005} \quad \text{or} \quad > z_{.005}$$

Since $\alpha = .01$ and $z_{.005} = 2.575$, the null hypothesis must be rejected if
$|Z| > 2.575$.

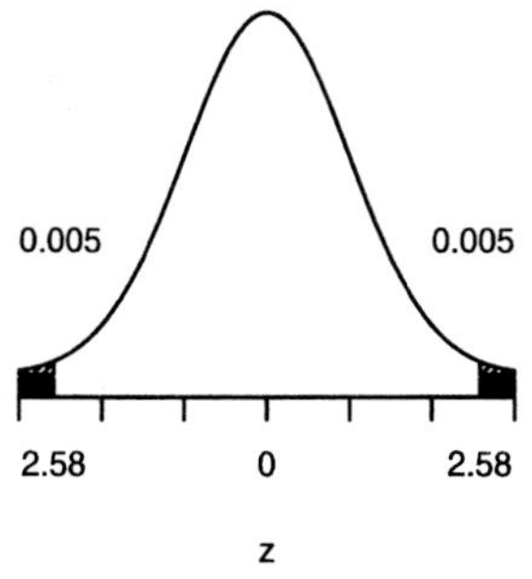

4. *Calculations:* $p_0 = .18$ $X = 24$, and $n = 160$ so

$$Z = \frac{24 - 160(.18)}{\sqrt{160(.18)(.82)}} = -.99$$

5. *Decision:* Since the observed value satisfies $|-.99| < 2.575$, we cannot reject the null hypothesis at the 1% level of significance.

10.53 1. *Null hypothesis* $H_0 : p = .10$
Alternative hypothesis $H_1 : p > .10$

2. *Level of significance:* $\alpha = 0.01$.

3. *Criterion:* Using a normal approximation for the binomial distribution, we reject the null hypothesis when

$$Z = \frac{X - np_0}{\sqrt{np_0(1-p_0)}} > z_{.01}.$$

Since $\alpha = .01$ and $z_{.01} = 2.33$, the null hypothesis must be rejected if $Z > 2.33$.

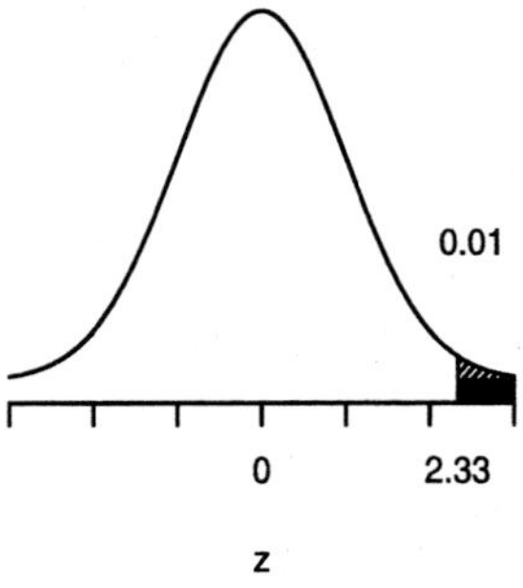

4. *Calculations:* $p_0 = .10$ $X = 16$, and $n = 100$ so

$$Z = \frac{16 - 100(.10)}{\sqrt{100(.10)(.90)}} = 2.00$$

5. *Decision:* Since the observed value is $2.00 < z_{.01} = 2.33$, we cannot reject the null hypothesis at the 1% level of significance.

10.55 1. *Null hypothesis* $H_0 : p_1 = p_2$

Alternative hypothesis $H_1 : p_1 > p_2$

2. *Level of significance:* $\alpha = 0.05$.

3. *Criterion:* We using the large sample statistic and reject the null hypothesis when

$$Z = \frac{X_1/n_1 - X_2/n_2}{\sqrt{\hat{p}(1-\hat{p})(\frac{1}{n_1} + \frac{1}{n_2})}} \quad \text{with } \hat{p} = \frac{X_1 + X_2}{n_1 + n_2}$$

is greater than $z_{.05} = 1.645$.

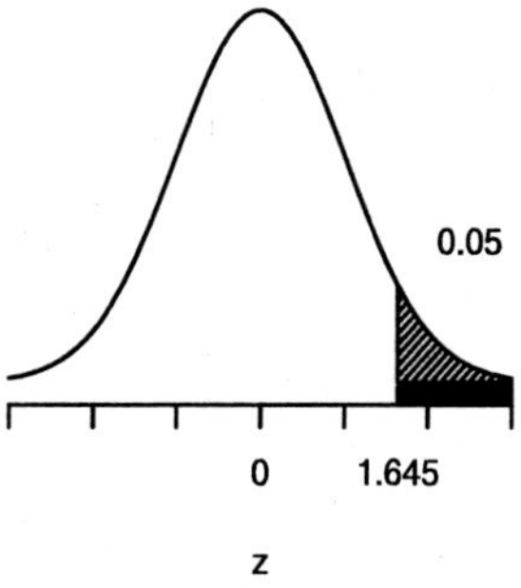

4. *Calculations:* In this case, $x_1 = 57$, $n_1 = 150$, $x_2 = 33$, $n_2 = 100$, and

$$\hat{p} = \frac{57 + 33}{150 + 100} = .36$$

Hence

$$Z = \frac{57/150 - 33/100}{\sqrt{(.36)(.64)(1/150\ +\ 1/100)}} = .807.$$

5. *Decision:* Since the observed value $.807 < z_{.05} = 1.645$, we cannot conclude at the 5% significance level that the first procedure is better than the second.

10.57 1. *Null hypothesis* $H_0 : p_A = p_B$

Alternative hypothesis $H_1 : p_A < p_B$

2. *Level of significance:* $\alpha = 0.05$.

3. *Criterion:* We using the large sample statistic and reject the null hypothesis when

$$Z = \frac{X_1/n_1 - X_2/n_2}{\sqrt{\hat{p}(1-\hat{p})(\frac{1}{n_1} + \frac{1}{n_2})}} \quad \text{with} \ \hat{p} = \frac{X_1 + X_2}{n_1 + n_2}$$

is less than $-z_{.05} = -1.645$.

4. *Calculations:* In this case, $x_1 = 11$, $n_1 = 50$, $x_2 = 19$, $n_2 = 50$, and

$$\hat{p} = \frac{11 + 19}{50 + 50} = .30$$

Hence

$$Z = \frac{11/50 - 19/50}{\sqrt{(.30)(.70)(1/50\ +\ 1/50)}} = -1.746.$$

5. *Decision:* Since the observed value $-1.746 < -z_{.05} = -1.645$, Thus, we reject the null hypothesis at the 5 % level and conclude that agent A is better than agent B.

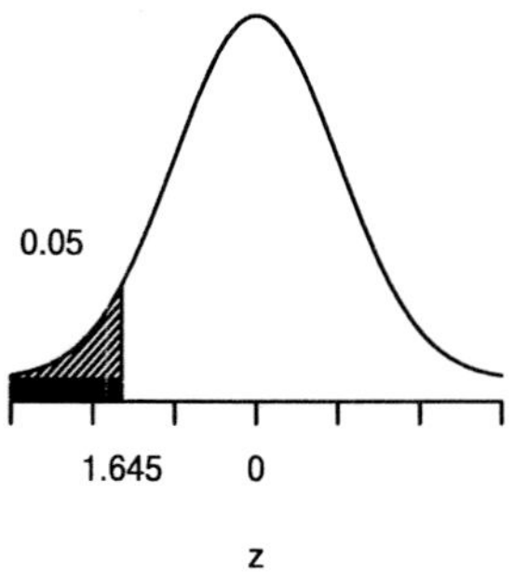

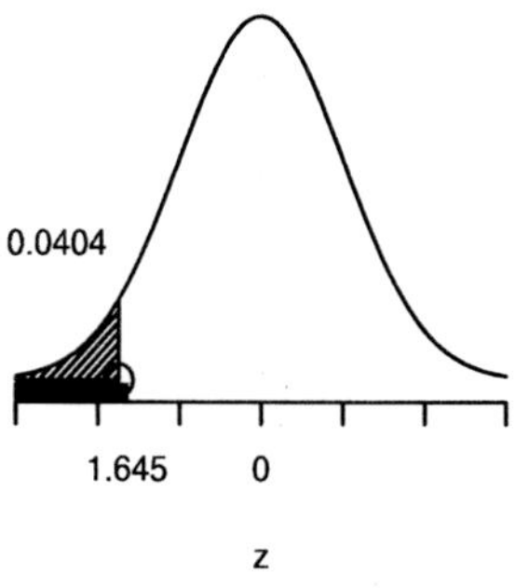

10.59 (a) We use the χ^2 statistic with 2 degrees of freedom to test the null hypothesis that the proportions of clogged cooling pipes at the three plants are equal against the alternative that they are not equal. Thus, we reject the null hypothesis at the 5% level when $\chi^2 > \chi^2_{.05} = 5.991$. In Table 10.15, the expected frequency and the contribution to the χ^2 statistic of each cell are given in the parentheses and brackets respectively.

Table 10.15. Exercise 10.59.

	Plant I	Plant II	Plant III	Total
Clogged	13 (13.33) [.01]	8 (13.33) [2.13]	19 (13.33) [2.41]	40
Unclogged	17 (16.67) [.01]	22 (16.67) [1.71]	11 (16.67) [1.93]	50
Total	30	30	30	90

The χ^2 statistic is

$$\begin{aligned}\chi^2 &= \frac{.33^2}{13.33} + \frac{5.33^2}{13.33} + \frac{5.67^2}{13.33} + \\ &\quad \frac{.33^2}{16.67} + \frac{5.33^2}{16.67} + \frac{5.67^2}{16.67} = 8.190.\end{aligned}$$

Thus, we reject the null hypothesis that the distribution of clogged cooling pipes is the same at all plants. The big contributions to χ^2 are due to low count of clogged pipes at Plant II and the high number at Plant III.

(b) The 95% confidence interval for the proportion of clogged pipes at the first plant is

$$\frac{13}{30} - 1.96\sqrt{\frac{(13/30)(17/30)}{30}} < p_1 < \frac{13}{30} + 1.96\sqrt{\frac{(13/30)(17/30)}{30}}$$

or $.256 < p_1 < .611$. At the second plant, the 95% confidence interval for the proportion of clogged pipes is

$$\frac{8}{30} - 1.96\sqrt{\frac{(8/30)(22/30)}{30}} < p_2 < \frac{8}{30} + 1.96\sqrt{\frac{(8/30)(22/30)}{30}}$$

or $.108 < p_2 < .425$. At the third plant, the 95% confidence interval for the proportion of clogged pipes is

$$\frac{19}{30} - 1.96\sqrt{\frac{(19/30)(11/30)}{30}} < p_3 < \frac{19}{30} + 1.96\sqrt{\frac{(19/30)(11/30)}{30}}$$

or $.461 < p_3 < .806$.

The plot of the confidence intervals is given below.

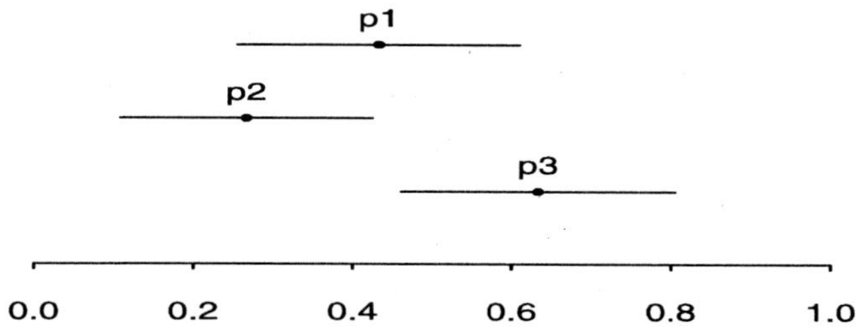

10.61 The data are summarized in the table:

Table 10.17. Exercise 10.61.

Number of failures	Number of days		Poisson prob. for $\lambda = 3.2$	Expected frequency	
0	9		.041	12.3	
1	43		.130	39.0	
2	64		.209	62.7	
3	62		.223	66.9	
4	42		.178	53.4	
5	36		.114	34.2	
6	22		.060	18.0	
7	14		.028	8.4	
8	6		.011	3.3	
9	2	8	.004	1.2	5.1
$\geq$ 10	0		.002	.6	

We use the χ^2 statistic with $9 - 1 = 8$ degrees of freedom to test the null hypothesis that the data are from a Poisson distribution with $\lambda = 3.2$. We reject the null hypothesis at the 5% level when $\chi^2 > \chi^2_{.05} = 15.507$. In this case,

$$\begin{aligned}\chi^2 &= \frac{3.3^2}{12.3} + \frac{4.0^2}{39.0} + \frac{1.3^2}{62.7} + \\ &\quad \frac{4.9^2}{66.9} + \frac{11.4^2}{53.4} + \frac{1.8^2}{34.2} + \\ &\quad \frac{4.0^2}{18.0} + \frac{5.6^2}{8.4} + \frac{2.9^2}{5.1} \\ &= 10.481.\end{aligned}$$

Thus, we cannot reject the null hypothesis.

10.63 We use the χ^2 statistic with 4 degrees of freedom to test the null hypothesis that appearance and X-ray inspection are independent. Thus, we reject the null hypothesis at the 5% level when $\chi^2 > \chi^2_{.05} = 9.488$. In Table 10.19, the expected frequency and the contribution to the χ^2 statistic of each cell are given in the parentheses and brackets respectively.

Table 10.19. Exercise 10.63.

		Appearance			
		Bad	Normal	Good	Total
X-ray	Bad	20 (8) [18.00]	7 (14) [3.50]	3 (8) [3.12]	30
	Normal	13 (21.33) [1.26]	51 (37.33) [5.00]	16 (21.33) [1.33]	80
	Good	7 (10.67) [1.26]	12 (18.67) [2.38]	21 (10.67) [10.01]	40
	Total	40	70	40	150

The χ^2 statistic is

$$\begin{aligned}\chi^2 &= \frac{12^2}{8} + \frac{7^2}{14} + \frac{5^2}{8} + \\ &\frac{8.33^2}{21.33} + \frac{13.67^2}{37.33} + \frac{5.33^2}{21.33} + \\ &\frac{3.67^2}{10.67} + \frac{6.67^2}{18.67} + \frac{10.33^2}{10.67} \\ &= 47.862.\end{aligned}$$

Thus, we reject the null hypothesis at the 5% level and conclude that there is dependence between appearance and the result of an x-ray inspection. The largest contributions to χ^2 come from the high counts in the Bad-Bad and Good-Good cells.

Chapter 11

REGRESSION ANALYSIS

11.1 (a) Use a random number table to select amount of additive for first run. Ignore 0, 6, 7, 8, and 9. Do the same for each of the four other runs but ignore amounts already selected.

(b) The experiment used the values 1, 2, 3, 4, and 5 but 8 is quite far beyond this range of values. Don't extrapolate beyond the range of experimental values because the model may change.

11.3 (a) The hand-drawn line in the scattergram in Figure 11.1 produces a prediction of 67 percent for the extraction efficiency when the extraction time is 35 minutes.

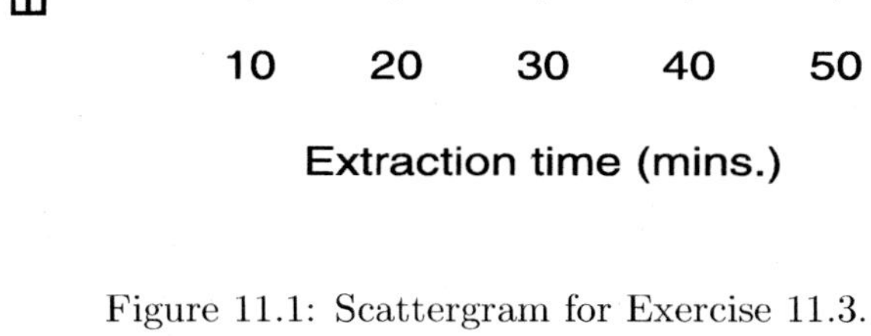

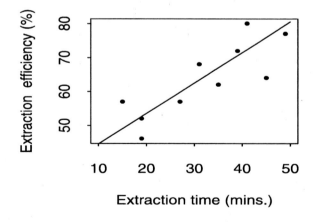

Figure 11.1: Scattergram for Exercise 11.3.

(b) In this example, n=10

$$\sum_{i=1}^{n} y_i = 635, \quad \sum_{i=1}^{n} x_i = 320, \quad \sum_{i=1}^{n} x_i^2 = 11,490, \quad \sum_{i=1}^{n} x_i y_i = 21,275$$

so,

$$
\begin{aligned}
S_{xx} &= \sum_{i=1}^{n} x_i^2 - (\sum_{i=1}^{n} x_i)^2/n = 11,490 - (320)^2/10 = 1250, \\
S_{xy} &= \sum_{i=1}^{n} x_i y_i - (\sum_{i=1}^{n} x_i)(\sum_{i=1}^{n} y_i)/n = 21,275 - 320(635)/10 = 955
\end{aligned}
$$

Consequently,

$$
\begin{aligned}
b &= \frac{S_{xy}}{S_{xx}} = \frac{955}{1250} = .764 \\
a &= \overline{y} - b\overline{x} = \frac{635}{10} - \frac{955}{1250}\frac{320}{10} = 39.052
\end{aligned}
$$

Thus, the equation for the least squares line is:

$$y = 39.052 + .764x$$

The prediction of the extraction efficiency when the extraction time is 35 minutes is

$$39.052 + (.764)(35) = 65.79 \text{ percent}$$

11.5 (a) Using the sums from Exercise 11.4b,

$$
\begin{aligned}
S_{xx} &= \sum_{i=1}^{n} x_i^2 - (\sum_{i=1}^{n} x_i)^2/n = 91 - (21)^2/6 = 17.5, \\
S_{yy} &= \sum_{i=1}^{n} y_i^2 - (\sum_{i=1}^{n} y_i)^2/n = 19855 - (311)^2/6 = 3,734.834, \\
S_{xy} &= \sum_{i=1}^{n} x_i y_i - (\sum_{i=1}^{n} x_i)(\sum_{i=1}^{n} y_i)/n = 1342 - 21(311)/6 = 253.5
\end{aligned}
$$

Thus,

$$s_e^2 = \frac{S_{xx}S_{yy} - (S_{xy})^2}{(n-2)S_{xx}} = \frac{(17.5)(3,734.834) - (253.5)^2}{(4)(17.5)} = 15.676.$$

The 95 percent confidence interval for β is given by

$$b \pm t_{\frac{\alpha}{2}}\, s_e \sqrt{\frac{1}{S_{xx}}}$$

or, since $t_{.025}$ for 4 degrees of freedom is 2.776,

$$14.486 \pm 2.776\sqrt{15.676}\sqrt{\frac{1}{17.5}}.$$

With 95 % confidence, β is between 11.86 and 17.11 thousandths of an inch per thousand pounds.

(b) The limits of prediction at x_0 are

$$a + bx_0 \pm t_{\frac{\alpha}{2}}\, s_e \sqrt{1 + \frac{1}{n} + \frac{(x_0 - \bar{x})^2}{S_{xx}}}$$

For Exercise 11.4, when the tensile force is 3.5 thousand pounds, this becomes

$$51.83 \pm 2.776\sqrt{15.676}\sqrt{1 + \frac{1}{6} + \frac{(3.5 - 3.5)^2}{17.5}}$$

The 95 % prediction interval for mean elogation, when $x = 3.5$, is from 39.96 to 63.71 10^{-3}in.

11.7 1. *Null hypothesis:* $\beta = 1.2$

Alternative hypothesis: $\beta < 1.2$.

2. *Level of significance:* $\alpha = 0.05$.

3. *Criterion:* Reject the null hypothesis if $t < -2.132$, where 2.132 is the value of $t_{0.05}$ for $6 - 2 = 4$ degrees of freedom , and t is given by the second formula of Theorem 11.1.

4. *Calculations:* We must first calculate S_{xx}, S_{xy}, S_{yy} and s_e. Since $\sum y_i^2 = 2001$, using the results of Exercise 11.4(a) gives

$$\begin{aligned} S_{xx} &= 304 - (36)^2/6 = 88 \\ S_{xy} &= 721 - (36)(107)/6 = 79 \\ S_{yy} &= 2001 - (107)^2/6 = 92.833 \end{aligned}$$

so that

$$s_e^2 = \frac{(88)(92.833) - (79)^2}{(4)(88)} = 5.478$$

Also, from Exercise 11.4(a), $b = .8977$ and $\beta = 1.2$ under the null hypothesis, and hence,

$$t = \frac{b - \beta}{s_e}\sqrt{S_{xx}} = \frac{.8977 - 1.2}{\sqrt{5.478}}\sqrt{88} = -1.212$$

5. *Decision:* Since -1.212 is greater than -2.132, we fail to reject the null hypothesis at the 0.05 level of significance.

11.9 (a) We calculate

	x	y	$x-\bar{x}$	$y-\bar{y}$	$(x-\bar{x})(y-\bar{y})$	$(x-\bar{x})^2$	$(y-\bar{y})^2$
	1	2	−2	−4	8	4	16
	2	5	−1	−1	1	1	1
	3	4	0	−2	0	0	4
	4	9	1	3	3	1	9
	5	10	2	4	8	4	16
Total	15	30	0	0	20	10	46

So $S_{xx} = 10$, $S_{yy} = 46$ and $S_{xy} = 20$. Also, $\bar{x} = 3$ and $\bar{y} = 6$. Thus,

$$b = \frac{100}{50} = 2$$
$$a = 6 - (2)(3) = 0$$

So, the least squares line is

$$\hat{y} = 0 + 2x$$

(b) When $x = 3.5$, the prediction is

$$\hat{y} = 2(3.5) = 7$$

11.11 We have to test the null hypothesis $H_0 : \beta = 1$ against the alternative $H_1 : \beta > 1$ at 0.05 level of significance. The t statistic is given by

$$t = \frac{b - \beta}{s_e} \sqrt{S_{xx}}$$

In our case, $S_{xx} = 10$ and $b = 2$. So,

$$t = \frac{(2 - 1)}{\sqrt{2}} \sqrt{10} = 2.236$$

Since $t_{.05}$ is 2.353 for 3 degrees of freedom, we cannot reject the null hypothesis $H_0 : \beta = 1$.

11.13 This calculation is the same as in 11.10b except that 1 is added under the square root and t is now $t_{.025}$ with 10 degrees of freedom. The 95 % prediction interval for the moisture content is

$$9.812 \pm 2.228\ (1.101) \sqrt{1 + \frac{1}{12} + \frac{(40 - 44.4)^2}{854.9167}}$$

or from 7.232 to 12.392 %. The 1/12 term is small compared to influence of s_e.

11.15 (a) We wish to minimize

$$\sum_{i=1}^{n}(y_i - \beta x_i)^2 = \sum_{i=1}^{n}(y_i - bx_i + bx_i - \beta x_i)^2$$
$$= \sum_{i=1}^{n}(y_i - bx_i)^2 + (b - \beta)^2 \sum_{i=1}^{n} x_i^2 + 2(\beta - b) \sum_{i=1}^{n}(y_i - bx_i)x_i$$

$$= \sum_{i=1}^{n}(y_i - bx_i)^2 + (b-\beta)^2 \sum_{i=1}^{n} x_i^2 \; + \; 0$$

by the definition of b. The first term does not depend on β and the second is a minimum for $\beta = b$, thus establishing the result.

(b) In Exercise 11.2, $\sum_{i=1}^{n} y_i x_i = 1342$ and $\sum_{i=1}^{n} x_i^2 = 91$, so that $b = 1342/91 = 14.747$. The previous slope estimate was 14.486. This estimate (14.747) is a little larger.

11.17 To calculate s_e^2, we need S_{yy}, which is equal to 21,401,588. Thus,

$$\begin{aligned} s_e^2 &= \frac{(215,065.88)(21,401,588) - (2,145,358.8)^2}{(6)(215,065.88)} \\ &= 144.89816, \end{aligned}$$

or

$$s_e = 12.0373.$$

Thus, the 95 percent limits are ($t_{.05}$ for 6 d.f. $= 1.943$)

$$22.90 \pm 1.943 \; (12.0373) \sqrt{\frac{1}{8} + \frac{(117.625)^2}{215,065.88}}$$

With 90 % confidence α is between 12.713 and 33.067.

11.19 (a) Instead of $x = (1,2,3,4,5,6,7)$, we will use $x = (-3,-2,-1,0,1,2,3)$. Then, $\bar{x} = 0$. Thus,

$$a = \sum_{i=1}^{n} y_i/n \quad \text{and} \quad b = \sum_{i=1}^{n} x_i y_i / \sum_{i=1}^{n} x_i^2$$

Since, $\sum_{i=1}^{n} y_i = 23.7$, $\sum_{i=1}^{n} x_i y_i = 19$, $\sum_{i=1}^{n} x_i^2 = 28$,

$$a = 23.7/7 = 3.386 \quad \text{and} \quad b = .679$$

Thus, the least squares line is $\hat{y} = 3.386 + .649x$. The 8'th year corresponds to $x = 4$, so the prediction is

$$3.386 + .679(4) = 6.102.$$

(b) We use $x = (-7,-5,-3,-1,1,3,5,7)$. Thus, $\sum_{i=1}^{n} y_i = 23.6$, $\sum_{i=1}^{n} x_i y_i = 40.6$, $\sum_{i=1}^{n} x_i^2 = 168$, so,

$$a = 23.6/8 = 2.95 \quad \text{and} \quad b = .24167$$

Thus, the least squares line is $\hat{y} = 2.95 + .24167x$. Since 2011 corresponds to $x = 10$, so the prediction for y is

$$\hat{y} = 2.95 + .24167(10) = 5.367$$

11.21 From Exercise 11.9 we form the following table

	x	y	$\hat{y} = 2x$	$(y - \hat{y})^2$	$(\bar{y} - \hat{y})^2$	$(y - \bar{y})^2$
	1	2	2	0	16	16
	2	5	4	1	4	1
	3	4	6	4	0	4
	4	9	8	1	4	9
	5	10	10	0	16	16
Total	15	30	30	6	40	46

Thus, the decomposition of the sum of squares is calculated as $46 = 6 + 40$, and

$$r^2 = 1 - 6/46 = 40/46 = 0.870$$

11.23 Using MINITAB we first obtain the least squares line.

(a)
```
The regression equation is
   height(nm) = 87.9 + 2.46 nanodiameter

Predictor         Coef  SE Coef     T      P
Constant         87.88    22.94  3.83  0.000
nanodiameter    2.4644   0.2573  9.58  0.000

S = 21.8953   R-Sq = 65.6%   R-Sq(adj) = 64.9%

Analysis of Variance
Source          DF     SS     MS      F      P
Regression       1  43965  43965  91.71  0.000
Residual Error  48  23011    479
Total           49  66976
```

(b) We present a the details to verify that $t = 9.66$ as in the output. We first calculate

$$S_{xx} = \sum_{i=1}^{50} x_i^2 - (\sum_{i=1}^{50} x_i)^2/50 = 397437 - (4417)^2 / 50 = 7239.22$$

1. *Null hypothesis* $H_0 : \beta = 0$

   Alternative hypothesis $H_1 : \beta \neq 0$

2. *Level of significance:* $\alpha = 0.05$.

3. *Criterion:* The test statistic is

$$t = \frac{\widehat{\beta}}{s_e}\sqrt{S_{xx}}$$

The alternative is two-sided so we reject the null hypothesis for $t < -t_{.025}$ or $t > t_{.025}$ where $t_{.025} = 2.011$ with 48 degrees of freedom. (We could use $z_{.025} = 1.96$ since the degrees of freedom are large.)

4. *Calculations:* From the computer output $s_e = 21.8953$ so

$$t = \frac{2.4644}{21.8953}\sqrt{7239.22} = 9.58$$

5. *Decision:* Because $9.58 > 2.011$, we reject the null hypothesis at the .05 level of significance. The P-value is 0 to several decimal places giving extremely strong support for rejecting the null hypothesis.

(c) At $x = 100$ nm we estimate the mean height $\alpha + \beta\,100$ as

Fit	SE Fit	95% CI
334.31	4.31	(325.65, 342.98)

or $325.65 < \alpha + \beta\,100 < 342.98$. We are 95 % confident that, when the mean width = 100 nm, the mean height is between 325.65 nm and 342.98 nm .

(d)

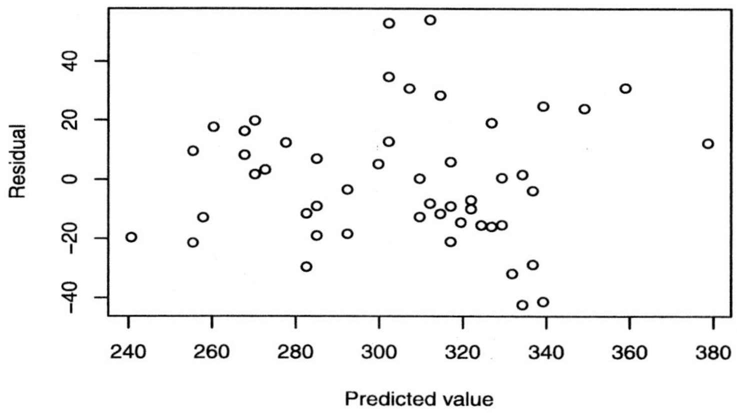

11.25 (a) The scattergram is given in Figure 11.2.

(b) Let $z_i = \log y_i$. Then, $\bar{x} = 10.5$, $\bar{z} = 5.476$, $S_{xx} = 157.5$, $S_{xz} = 9.507$. Thus,

$$b = \frac{S_{xz}}{S_{xx}} = 0.0604 \text{ and } a = \bar{z} - b\bar{x} = 4.842$$

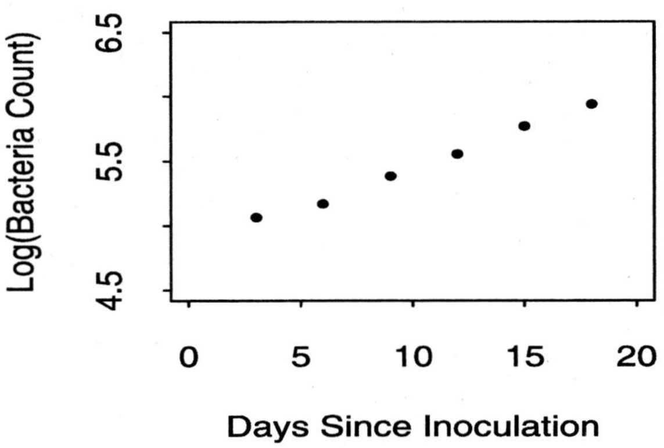

Figure 11.2: Scattergram for Exercise 11.25

So, the fit is

$$\hat{y} = 10^{4.842 + 0.0604x} = 69{,}502.2(1.149)^x$$

(c) Thus, the prediction when $x = 20$, is

$$10^{4.842 + 0.0604(20)} = 1{,}122{,}018$$

11.27 The equation obtained in Exercise 11.26 is

$$\hat{y} = 10^{1.452 + .0000666x} = (10^{1.452})(10^{.0000666x}) = 28.314e^{.0001534x}$$

When $x = 3000$, $\hat{y} = 44.86$.

11.29 The data are

Altitude	Dose Rate
50	28
450	30
780	32
1,200	36
4,400	51
4,800	58
5,300	69

Since the model is

$$y = \exp\left[e^{\alpha x + \beta}\right]$$

we have

$$\ln\,(\ln y) \;=\; \alpha x \;+\; \beta$$

Thus, the data to which we are fitting a least squares line are

Altitude	ln (ln (Dose Rate))
50	1.2036
450	1.2241
780	1.2429
1,200	1.2703
4,400	1.3691
4,800	1.4013
5,300	1.4432

Let $z \;=\; \ln(\ln y)$, then

$$n \;=\; 7, \quad \sum x \;=\; 16,980 \quad \sum x^2 \;=\; 72,743,400$$

$$\sum z \;=\; 9.1605, \quad \sum z^2 \;=\; 12.042, \quad \sum xz \;=\; 23511.29$$

Hence

$$S_{xx} \;=\; 72,743,400 \;-\; (16,980)^2/7 \;=\; 31,554,771$$

$$S_{xz} \;=\; 23511.29 \;-\; (16,980)(9.1605)/7 \;=\; 1290.5343$$

Consequently, we get

$$b \;=\; \frac{S_{xz}}{S_{xx}} \;=\; 0.0000409, \quad \text{and} \quad a \;=\; \bar{z} \;-\; b\bar{x} \;=\; 1.2094$$

Thus, the fitted equation is $\hat{y} = \exp(\exp(0.00004x + 1.2))$.

11.31 $y = 3 - 3e^{-\alpha x}$ can be rewritten as

$$\ln(1 \;-\; y/3) \;=\; -\,\alpha x$$

Let $y_i' = \ln(1 - y_i/3)$ and $x_i' = -x_i$. Thus, using the no-intercept least square fit

$$\alpha \;=\; \frac{\sum x' y'}{\sum x'^2} \;=\; \frac{264.3584}{1100} \;=\; .240$$

11.33 (a) Fitting a straight line to the data gives

$$\hat{y} \;=\; 10.4778 \;-\; 0.38334x$$

where

$$SSE_1 = \sum (y - \hat{y})^2 = 11.906,$$

thereby giving

$$\hat{\sigma_1}^2 = 11.906/9 = 1.32289.$$

The t statistic for the null hypothesis $\beta_1 = 0$ is given by

$$t = \frac{-0.38334 - 0}{\sqrt{11.906/7}} \sqrt{\frac{9.60}{9}} = -2.28.$$

(b) In this case, $\hat{\sigma_2}^2 = \sum (y - \hat{y})^2/\text{d.f.} = 1.602/6 = 0.2670$ and $\hat{y} = 12.1848 - 1.8465 + 1.829x^2$. Thus, the F statistic is given by

$$\frac{11.906 - 1.602}{0.2670} = \frac{10.304}{0.2670} = 38.59176$$

The critical value for $F_{.05}$ with 1 and 6 degrees of freedom is 5.99 so the test is not significant at the 5 percent level.

11.35 We wish to minimize

$$D \equiv D(\beta_0, \beta_1, \beta_2) = \sum_{i=1}^{n} [y_i - (\beta_0 + \beta_1 x_{1i} + \beta_2 x_{2i})]^2$$

Taking derivatives

$$\frac{\partial D}{\partial \beta_0} = -2 \sum_{i=1}^{n} [y_i - (\beta_0 + \beta_1 x_{1i} + \beta_2 x_{2i})]$$

$$\frac{\partial D}{\partial \beta_1} = -2 \sum_{i=1}^{n} [y_i - (\beta_0 + \beta_1 x_{1i} + \beta_2 x_{2i})]\, x_{1i}$$

$$\frac{\partial D}{\partial \beta_2} = -2 \sum_{i=1}^{n} [y_i - (\beta_0 + \beta_1 x_{1i} + \beta_2 x_{2i})]\, x_{2i}$$

Setting these to zero gives

$$\sum y = nb_0 + b_1 \sum x_1 + b_2 \sum x_2$$

$$\sum x_1 y = b_0 \sum x_1 + b_1 \sum x_1^2 + b_2 \sum x_1 x_2$$

$$\sum x_2 y = b_0 \sum x_2 + b_1 \sum x_1 x_2 + b_3 \sum x_2^2$$

11.37 Using $\hat{y} = 161.34 + 32.97x_1 - .086x_2$ gives $\hat{y} = 64.0885$ for $x_1 = .05$ and $x_2 = 1150$.

11.39 The system of linear equations to be solved is

$$\begin{aligned} 84.6 &= 10b_0 + 15.3b_1 + 939b_2 \\ 132.27 &= 15.3b_0 + 29.85b_1 + 1{,}458.9b_2 \\ 8{,}320.2 &= 939b_0 + 1{,}458.9b_1 + 94{,}131b_2 \end{aligned}$$

Solving gives the fitted line

$$\hat{y} = 2.266 + 0.225x_1 + 0.0623x_2$$

When $x_1 = 2.2$ and $x_2 = 90$, the predicted value is 8.368.

11.41

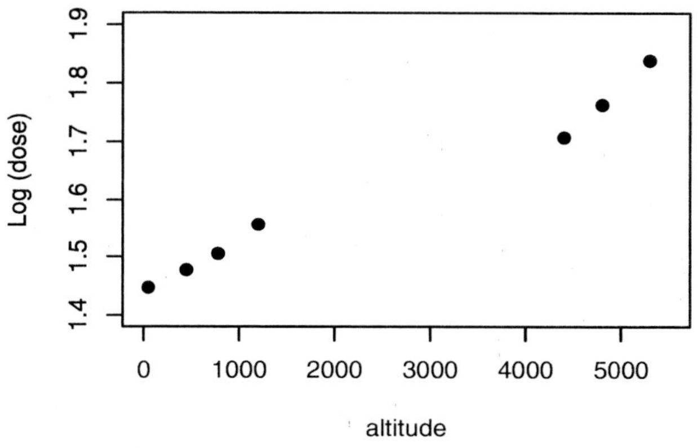


```
The regression equation is
Log(dose) = 1.45 +0.000067 altitude

Predictor         Coef      SE Coef          T        P
Constant       1.45203      0.01475      98.41    0.000
altitude    0.00006663   0.00000458      14.56    0.000

S = 0.02571       R-Sq = 97.7%      R-Sq(adj) = 97.2%

Analysis of Variance

Source            DF          SS          MS         F        P
Regression         1     0.14008     0.14008    211.91    0.000
Residual Error     5     0.00331     0.00066
Total              6     0.14339
```

11.43 (a) The residual plots are not too bad for $n = 15$. There is hint that the two smallest residuals are a little small to meet the normal assumption.

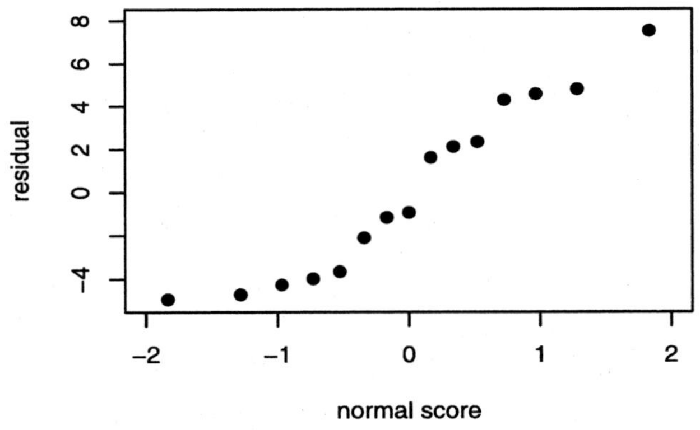

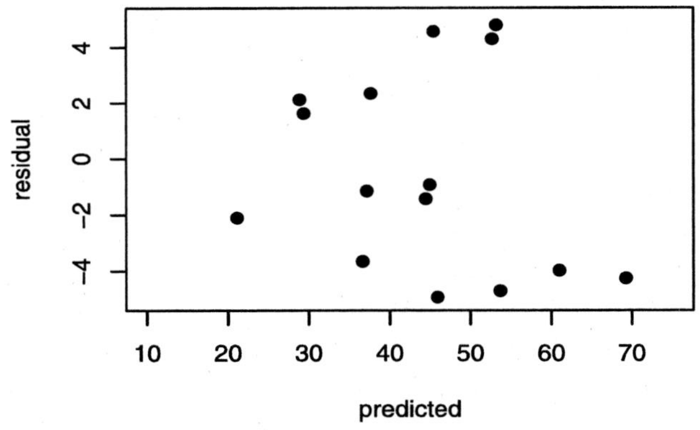

(b)
```
The regression equation is
hardness = 161 + 33.0 Cu - 0.0855 temp
```

Predictor	Coef	SE Coef	T	P
Constant	161.34	11.43	14.11	0.000
Cu	32.97	16.75	1.97	0.081
temp	-0.085500	0.009788	-8.74	0.000

```
S = 3.791        R-Sq = 89.9%      R-Sq(adj) = 87.7%
```

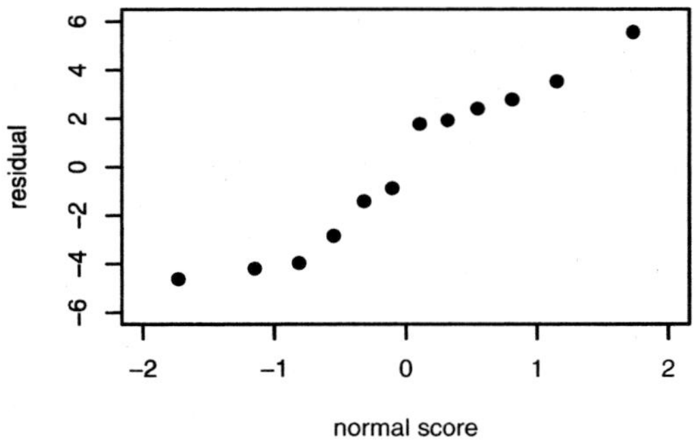

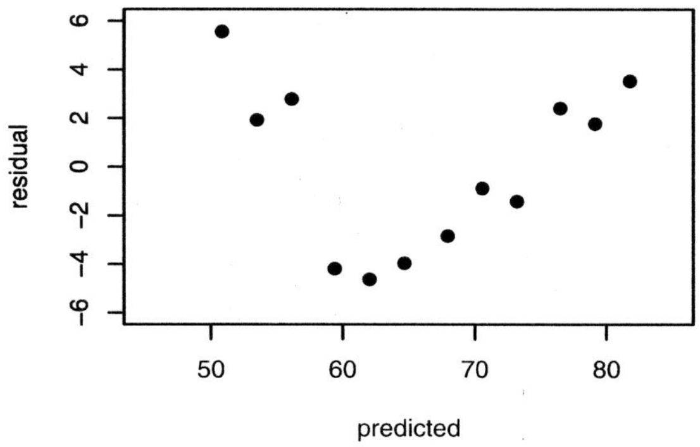

11.45 In the plot of the residual versus the time order there is a strong increasing pattern indicating a dependence over time. This violates the assumption of independent errors.

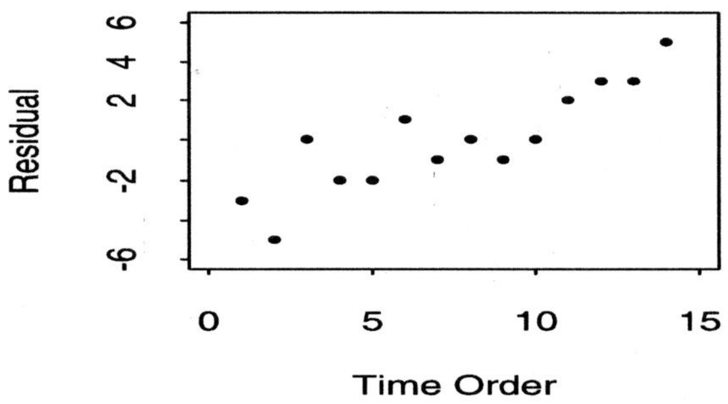

11.47 Ordering beer will not increase the number of weddings. There is a lurking variable here, the population size of cities. Large cities will likely have large values for both variables and small cities are likely to have small values of both variables.

11.49 (a) We first calculate

$$\sum x_i^2 = 532,000, \quad \sum x_i = 2,000, \quad \sum y_i^2 = 9.1097$$

$$\sum y_i = 8.35, \quad \sum x_i y_i = 2175.4$$

and

$$\begin{aligned} S_{xx} &= 532000 - (2000)^2/10 = 132,000 \\ S_{yy} &= 9.1097 - (8.35)^2/10 = 2.13745 \end{aligned}$$

$$S_{xy} = 2175.4 - (8.35)(2000)/10 = 505.4$$

Thus,

$$r = \frac{505.4}{\sqrt{(132000)(2.13745)}} = 0.9515.$$

(b) The statistic for testing $\rho = 0$, when the data are from a bivariate normal population is

$$Z = \frac{\sqrt{n - 3}}{2} \ln\left(\frac{1 + r}{1 - r}\right) = \frac{\sqrt{7}}{2} \ln\left(\frac{1 + .9515}{1 - .9515}\right) = 4.89$$

The critical value is 1.96 for a two-sided test with $\alpha = 0.05$, so the null hypothesis that $\rho = 0$ is rejected.

11.51 The statistic for testing the hypothesis that $\rho = 0$,when the data are from a bivariate normal population is

$$Z = \frac{\sqrt{n-3}}{2} \ln\left(\frac{1+r}{1-r}\right) = \frac{\sqrt{21}}{2} \ln\left(\frac{1+.809}{1-.809}\right) = 5.151$$

The critical value for $\alpha = 0.01$ is 2.56 , so the null hypothesis is rejected.

11.53 The required sums are

$$\sum x = 533 , \quad \sum x^2 = 24529 , \quad \sum y = 132$$

$$\sum y^2 = 1526 , \quad \sum xy = 6093$$

$$S_{xx} = 24529 - (533)^2/12 = 854.9167$$

$$S_{yy} = 1526 - (132)^2/12 = 74$$

$$S_{xy} = 6093 - (132)(533)/12 = 230$$

Thus,

$$r = \frac{230}{\sqrt{(854.9167)(74)}} = 0.914$$

$$\mathcal{Z} = \frac{1}{2}\ln\left(\frac{1+r}{1-r}\right) = \frac{1}{2}\ln\left(\frac{1+.914}{1-.914}\right) = 1.554$$

Since $z_{0.025} = 1.96$, the formula

$$\mathcal{Z} - \frac{z_{0.025}}{\sqrt{n-3}} < \mu_{\mathcal{Z}} < \mathcal{Z} + \frac{z_{0.025}}{\sqrt{n-3}}$$

gives a 95 percent confidence interval from 0.9007 to 2.207 for $\mu_{\mathcal{Z}}$. With 95 % confidence ρ is between 0.7166 and 0.9761.

11.55 The required sums are

$$\sum x = 4417, \sum x^2 = 397437, \sum y = 15279.0$$

$$\sum y^2 = 4735933, \sum xy = 1367587$$

Thus,

$$\begin{aligned} S_{xx} &= 397473 - (4417)^2/50 = 7239.22 \\ S_{yy} &= 473933 - (15279)^2/50 = 66976.2 \\ S_{xy} &= 1367587 - (4417)(15279)/50 = 17840.1 \end{aligned}$$

Thus,

$$r = \frac{S_{xy}}{\sqrt{S_{xx}S_{yy}}} = 0.810$$

Alternatively, MINITAB gives

```
Correlations: nanodiameter, height(nm)


Pearson correlation of nanodiameter and height(nm) = 0.810
P-Value = 0.000
```

11.57 We use

$$\begin{aligned} Z &= \frac{\sqrt{n-3}}{2} \ln \left[\frac{(1+r)(1-\rho_0)}{(1-r)(1+\rho_0)} \right] \\ &= \frac{\sqrt{22}}{2} \ln \left[\frac{(0.42)(1.40)}{(1.58)(0.60)} \right] = -1.120 \end{aligned}$$

The one-sided critical value at the 0.05 level is about -1.645, so the null hypothesis cannot be rejected.

11.59 (a)

$$f(x,y) = \frac{1}{2\pi\sigma\sigma_1} \exp\left[-\left(\frac{[y-(\alpha+\beta x)]^2}{2\sigma^2} + \frac{(x-\mu_1)^2}{2\sigma_1^2} \right) \right]$$

Rearranging the exponent and making a perfect square in x we obtain,

$$f(x,y) = \frac{1}{\sqrt{2\pi}\sqrt{\sigma^2+\sigma_1^2\beta^2}} \exp\left[-\frac{[y-(\alpha+\beta\mu_1)]^2}{2(\sigma^2+\sigma_1^2\beta^2)} \right] \times$$
$$\frac{\sqrt{\sigma^2+\sigma_1^2\beta^2}}{\sqrt{2\pi}\sigma\sigma_1} \exp\left[-\frac{(\sigma^2+\sigma_1^2\beta^2)}{2\sigma^2\sigma_1^2} \left(x + \frac{\sigma_1^2\alpha\beta - \sigma_1^2\beta y - \sigma^2\mu_1}{\sigma^2+\sigma_1^2\beta^2} \right)^2 \right]$$

Now, the marginal density of Y is given by

$$f_2(y) = \int_{-\infty}^{\infty} f(x,y)dx$$

But the first term in the integrand, being independent of x, comes out of the integral and the second term, being a normal density, integrates to 1. So,

$$f_2(y) = \frac{1}{\sqrt{2\pi}\sqrt{\sigma^2+\sigma_1^2\beta_1^2}} \exp\left[-\frac{[y-(\alpha+\beta\mu_1)]^2}{2(\sigma^2+\sigma_1^2\beta_1^2)}\right].$$

This is just a normal density with mean $\alpha+\beta\mu_1$ and variance $\sigma^2+\sigma_1^2\beta^2$. Thus,

$$\mu_2 = \alpha+\beta\mu_1 \text{ and } \sigma_2^2 = \sigma^2+\sigma_1^2\beta^2.$$

A far simpler way to show this is to use the formulas

$$E(Y) = E(E(Y|X)) \text{ and } V(Y) = V(E[Y|X]) + E(V[Y|X]).$$

Now $E(Y|X) = \alpha+\beta X$. So,

$$E(Y) = E(E(Y|X)) = \alpha+\beta E(X) = \alpha+\beta\mu_1.$$

Since $V(Y|X) = \sigma^2$, a constant,

$$V(Y) = V(\alpha+\beta X) + E(\sigma^2) = \beta^2 V(X) + \sigma^2 = \beta^2\sigma_1^2 + \sigma^2.$$

(b) We start with

$$f(x,y) = \frac{1}{2\pi\sigma\sigma_1} \exp\left[-\left(\frac{[y-(\alpha+\beta x)]^2}{2\sigma^2} + \frac{(x-\mu_1)^2}{2\sigma_1^2}\right)\right]$$

Substituting $\alpha = \mu_2 - \beta\mu_1$, $\sigma^2 = \sigma_2^2 - \beta^2\sigma_1^2$ in the formula, and observing $\rho^2 = 1-\sigma^2/\sigma_2^2$, we obtain,

$$f(x,y) = \frac{1}{2\pi\sigma_1\sigma_2\sqrt{1-\rho^2}} \times$$

$$\exp\left[-\left(\left(\frac{x-\mu_1}{\sigma_1}\right)^2 - 2\rho\left(\frac{x-\mu_1}{\sigma_1}\right)\left(\frac{y-\mu_2}{\sigma_2}\right) + \left(\frac{y-\mu_2}{\sigma_2}\right)^2\right)/2(1-\rho^2)\right]$$

11.61 The table gives y, $\hat{y}$ and the residuals $y-\hat{y}$

y	$\hat{y}$	$y - \hat{y}$	y	$\hat{y}$	$y - \hat{y}$
38	47.24	−9.24	31	29.69	1.31
40	55.06	−15.06	35	37.51	−2.51
85	62.89	22.11	42	45.34	−3.34
59	70.71	−11.71	59	53.16	5.84
40	38.46	1.54	18	20.91	−2.91
60	46.29	13.71	34	28.74	5.26
68	54.11	13.89	29	36.56	−7.56
53	61.94	−8.94	42	44.39	−2.39

(a) $\sum(y - \bar{y})^2 = 4318.44$

(b) $\sum(y - \hat{y})^2 = 1553.81$

(c) $r = \sqrt{1 - (1553.81)/(4318.44)} = 0.80$.

11.63 The required sums are

$$\sum x = 5180, \sum x^2 = 1276454, \sum y = 4668.0$$

$$\sum y^2 = 923838, \sum xy = 1008732$$

Thus,

$$S_{xx} = 1276454 - (5180)^2/30 = 382041$$
$$S_{yy} = 923838 - (4668)^2/30 = 197497$$
$$S_{xy} = 1008732 - (5180)(4668)/30 = 202724$$

Thus,

$$r = \frac{S_{xy}}{\sqrt{S_{xx}S_{yy}}} = 0.738$$

Alternatively, MINITAB gives

```
Pearson correlation of x and y = 0.738
P-Value = 0.000
```

11.65 (a) We calculate

	x	y	$x-\bar{x}$	$y-\bar{y}$	$(x-\bar{x})(y-\bar{y})$	$(x-\bar{x})^2$	$(y-\bar{y})^2$
	1	10	−2	−40	80	4	1600
	2	40	−1	−10	10	1	100
	3	30	0	−20	0	0	400
	4	80	1	30	30	1	900
	5	90	2	40	80	4	1600
Total	15	250	0	0	200	10	4600

So $S_{xx} = 10$, $S_{yy} = 4600$ and $S_{xy} = 200$.

$$b = S_{xy}/S_{xx} = 200/10 = 20, \qquad a = \overline{y} - b\overline{x} = 50 - 20(3) = -10$$

Thus,

$$\hat{y} = -10 + 20x.$$

(b) At $x = 4.5$, $\hat{y} = -10 + 20(4.5) = 80$.

(c) $x = 7$ is outside the experimental region and the straight line relationship may not hold.

11.67 1. *Null hypothesis:* $\beta = 5$
Alternative hypothesis: $\beta > 5$.

2. *Level of significance:* $\alpha = 0.05$.

3. *Criterion:* Reject the null hypothesis if $t > 2.353$, where 2.353 is the value of $t_{0.05}$ for $5 - 2 = 3$ degrees of freedom , and t is given by the second formula of Theorem 11.1.

4. *Calculations:* From Exercise 11.65,$n = 5$, $S_{xx} = 10$, $b = 20$ and $t_{0.05} = 2.353$, and $s_e^2 = 200$. The value of the t statistic is

$$t = \frac{b-5}{s_e}\sqrt{S_{xx}} = \frac{15}{14.142}\sqrt{10} = 3.354.$$

Thus, at level $\alpha = 0.05$, $H_0 : \beta = 5$ is rejected in favor of the alternative.

5. *Decision:* Since 3.354 is greater than 2.353, we reject the null hypothesis $H_0 : \beta = 5$ in favor of the alternative at the 0.05 level of significance. The $P-$ value is less than 0.025.

11.69 (a) $n = 5$ and

$$\overline{x} = \frac{15}{5} = 3 \quad \overline{y} = \frac{75}{5} = 15$$

so,

$$\begin{aligned} S_{xx} &= \sum_{i=1}^{n}(x_i - \overline{x})^2 = (-2)^2 + (-1)^2 + 0^2 + 1^2 + 2^2 = 10, \\ S_{xy} &= \sum_{i=1}^{n}(x_i - \overline{x})(y_i - \overline{y}) = (-2)(4) + (-1)(2) + (0)(-1) + (1)(-2) + (2)(-3) = -18 \end{aligned}$$

Consequently,

$$\begin{aligned} b &= \frac{S_{xy}}{S_{xx}} = \frac{-18}{10} = -1.8 \\ a &= \overline{y} - b\overline{x} = 15 - (-1.8)(3) = 20.4 \end{aligned}$$

Thus, the equation for the least squares line is:

$$y = 20.4 - 1.8x$$

(b) 1. *Null hypothesis:* $\beta = 0$

Alternative hypothesis: $\beta \neq 0$.

2. *Level of significance:* $\alpha = 0.10$.

3. *Criterion:* Reject the null hypothesis if $|t| > 2.353$, where 2.353 is the value of $t_{0.05}$ for $5 - 2 = 3$ degrees of freedom , and t is given by the second formula of Theorem 11.1.

4. *Calculations:* To obtain s_e^2, we first need

$$S_{yy} = \sum_{i=1}^{n}(y_i - \overline{y})^2 = (4)^2 + (2)^2 + (-1)^2 + (-2)^2 + (-3)^2 = 34$$

so that

$$s_e^2 = \frac{S_{yy} - S_{xy}^2 / S_{xx}}{n-2} = \frac{34 - (-18)^2/10}{3} = .5333$$

Also, from Part (a), $b = -1.8$ so the value of the t statistic is

$$t = \frac{b-0}{s_e}\sqrt{S_{xx}} = \frac{-1.8}{\sqrt{.5333}}\sqrt{10} = -7.794$$

5. *Decision:* Since -7.794 is less than -2.353, reject the null hypothesis at the 0.10 level of significance. The $P-$ value is less than 0.005.

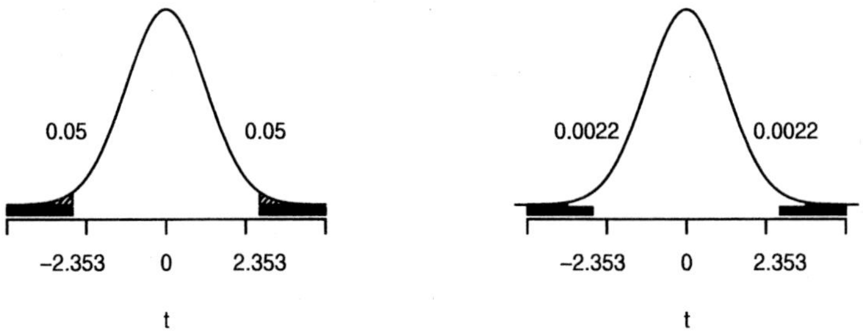

(c) The estimate of the mean NOx at $x = 9$ is

$$\hat{y} = a + b(9) = 20.4 - 1.8(9) = 4.2$$

and the 99 percent confidence interval is

$$a + b(9) \pm t_{.005}\, s_e \sqrt{\frac{1}{n} + \frac{(9-\overline{x})^2}{S_{xx}}} = 4.2 \pm 5.841\sqrt{.5333}\sqrt{\frac{1}{5} + \frac{(9-3)^2}{10}} = 4.2 + 8.315$$

So the interval is from -4.115 to 12.515 but we could use 0 as lower limit.

(d) The experiment used the values 1, 2 3, 4, and 5 but the value $x = 9$ for the predictor variable is quite far from these values. Don't extrpolate beyond the range of the experiment. The model may change in the region near 9.

11.71 We have

$$\begin{aligned} S_{xx} &= \sum_{i=1}^{n}(x_i - \overline{x})^2 = (-2)^2 + (-1)^2 + 0^2 + 1^2 + 2^2 = 10, \\ S_{xy} &= \sum_{i=1}^{n}(x_i - \overline{x})(y_i - \overline{y}) = (-2)(4) + (-1)(2) + (0)(-1) + (1)(-2) + (2)(-3) = -18 \\ S_{yy} &= \sum_{i=1}^{n}(y_i - \overline{y})^2 = (4)^2 + (2)^2 + (-1)^2 + (-2)^2 + (-3)^2 = 34 \end{aligned}$$

so $r = S_{xy}\sqrt{S_{xx}\,S_{yy}} = -18\,/\sqrt{10 \cdot 34} = -.9762$. The proportion of variance of NOx explained by the additive is $r^2 = (-.7 = 9762)^2 = .9529$.

11.73 The variance estimate is $s_e = .4154$ by (c) of 11.68. With $t_{.025} = 2.306$, a 95 percent confidence interval for α is given by

$$.619 \pm 2.306(.4154)\sqrt{\frac{1}{10} + \frac{3.5^2}{124.375}} = .619 \pm .427$$

With 95 % confidence, *alpha* is between 192 and 1.046.

11.75 The ideal gas law is $pV^{\gamma} = C$. Taking logs of both sides gives

$$\ln p + \gamma \ln V = C.$$

Let $y_i = \ln p_i$ and $x_i = -\ln V_i$. Then, $\bar{x} = -2.815$, $\bar{y} = 4.530$, $S_{xx} = 3.3068$, $S_{xy} = 4.9573$. So,

$$\begin{aligned} \gamma &= \frac{S_{xy}}{S_{xx}} = 1.499 \\ C &= \overline{y} - b\overline{x} = 8.760 \end{aligned}$$

11.77 For the equation $I = 1 - e^{-\epsilon/\tau}$,

$$\ln(1 - I) = -\frac{1}{\tau}t.$$

Let $y_i = \ln(1 - I_i)$. Notice that this equation does not have an intercept, so use

$$b = \frac{\sum_{i=1}^{n} y_i t_i}{\sum_{i=1}^{n} t_i^2}.$$

$$\sum_{i=1}^{n} y_i t_i = -2.1793, \quad \sum_{i=1}^{n} t_i^2 = 2.04.$$

So

$$b = \frac{-2.1793}{2.04} = -1.068 \quad \text{and} \quad \tau = -\frac{1}{b} = 0.936.$$

11.79 (a) From Exercise 11.9, $S_{xx} = 10$, $S_{xy} = 20$, $S_{yy} = 46$ and $\hat{y} = 0 + 2x$. Also $s_e^2 = 2.00$. The prediction for the mean CPU time at $x = 3.0$ is $\hat{y} = 0 + 2(3.0) = 6$. The 95 percent confidence interval is given by

$$6 \pm 3.182\sqrt{2.00}\sqrt{\frac{1}{5} + \frac{(3-3)^2}{10}}.$$

With 95 % confidence the mean CPU time, when $x = 3.0$, is between 3.99 to 8.01 hours.

(b) The 95 percent limits of prediction for a single future day is

$$6 \pm 3.182\sqrt{2.00}\sqrt{1 + \frac{1}{5} + \frac{(3-3)^2}{10}}.$$

The 95% prediction interval for a future day with $x = 3.04$ is from 1.07 to 10.93 hours.

11.81 $r_1 = 0.41$, $r_1^2 = 0.1681$, $r_2 = 0.29$, $r_2^2 = 0.0841$. Thus, in the first relationship 16.81 percent of the variation is explained by a linear relationship and in the second relationship 8.41 percent of the variation is explained by a linear relationship. With regard to variance, the linear relationship in the first case is relatively about twice as strong.

11.83 (a) Since the transformation $\mathcal{Z} = (1/2)\ln(1 + r/1 - r) = 1.045$ and $z_{0.025} = 1.96$, the 95% confidence interval for $\mu_{\mathcal{Z}}$ is $\mathcal{Z} \pm 1.96/\sqrt{n-3}$ or from 0.4796 to 1.6112. With 95 % confidence, ρ is between 0.446 and 0.923 .

(b) Proceeding as in part (a), $\mathcal{Z} = -0.725$. The interval for $\mu_{\mathcal{Z}}$ is from -1.089 to -0.361. With 95 % confidence, ρ is between -0.797 and -0.346.

(c) $\mathcal{Z} = 0.1717$. Thus, the interval for $\mu_{\mathcal{Z}}$ is from -0.1748 to 0.5181. With 95 % confidence, ρ is between -0.173 and 0.476.

11.85 (a) We have $n = 26$

$$\begin{aligned} \textstyle\sum_{i=1}^{n} x_i &= 102, & \textstyle\sum_{i=1}^{n} x_i^2 &= 428, \\ \textstyle\sum_{i=1}^{n} y_i &= 47.057, & \textstyle\sum_{i=1}^{n} x_i y_i &= 197.99. \end{aligned}$$

so

$$S_{xx} = \sum_{i=1}^{n} x_i^2 - (\sum_{i=1}^{n} x_i)^2/n = 428 - (102)^2/26 = 27.8462$$
$$S_{xy} = \sum_{i=1}^{n} x_i y_i - (\sum_{i=1}^{n} x_i)(\sum_{i=1}^{n} y_i)/n = 197.99 - 102(47.057)/26 = 13.3817$$

Consequently,

$$b = \frac{S_{xy}}{S_{xx}} = \frac{13.382}{27.846)} = .048$$
$$a = \overline{y} - b\overline{x} = \frac{47.057}{26} - \frac{13.3817}{27.8462}\frac{102}{26} = -.075$$

So, the least squares line is

$$\hat{y} = -0.075 + 0.48x.$$

(b) In addition to the sums $S_{xx} = 27.846$ and $S_{xy} = 13.382$ from Part (a), we calculate

$$S_{yy} = \sum_{i=1}^{n} y_i^2 - (\sum_{i=1}^{n} y_i)^2/n = 91.794 - (47.057)^2/26 = 6.626,$$

Thus,

$$s_e^2 = \frac{S_{xx}S_{yy} - (S_{xy})^2}{(n-2)S_{xx}}$$
$$= \frac{(27.846)(6.626) - (13.382)^2}{(24)(27.846)} = .0081246$$

The 95 percent confidence interval is given by

$$b \pm t_{\frac{\alpha}{2}} s_e \sqrt{\frac{1}{S_{xx}}}$$

or, since $t_{.025}$ for 24 degrees of freedom is 2.064,

$$0.48 \pm 2.064\sqrt{.0081246}\sqrt{\frac{1}{27.846}}$$

With 95% confidence, the slope(or charge on the electron), is between .44 to .52 charge$\times 10^9$.

(c) We have to test the null hypothesis $H_0 : \alpha = 0$ against the alternative $H_1 : \alpha \neq 0$ at 0.05 level of significance. The t statistic is given by

$$t = \frac{a - \alpha}{s_e} \sqrt{\frac{nS_{xx}}{S_{xx} + n\bar{x}^2}}.$$

In our case, $S_{xx} = 27.846$ and $a = -0.07511$. So,

$$t = \frac{(-0.07511 - 0)}{\sqrt{.0081246}} \sqrt{\frac{26(27.846)}{27.846 + 26(3.923)^2}} = -1.084.$$

Since $t_{.025}$ is 2.064 for 24 degrees of freedom, we cannot reject the null hypothesis $H_0 : \alpha = 0$.

(d) The residual verus fitted value plot does not really have enough residuals to reveal a strong pattern and we proceed as if the assumptions of the model are not violated.

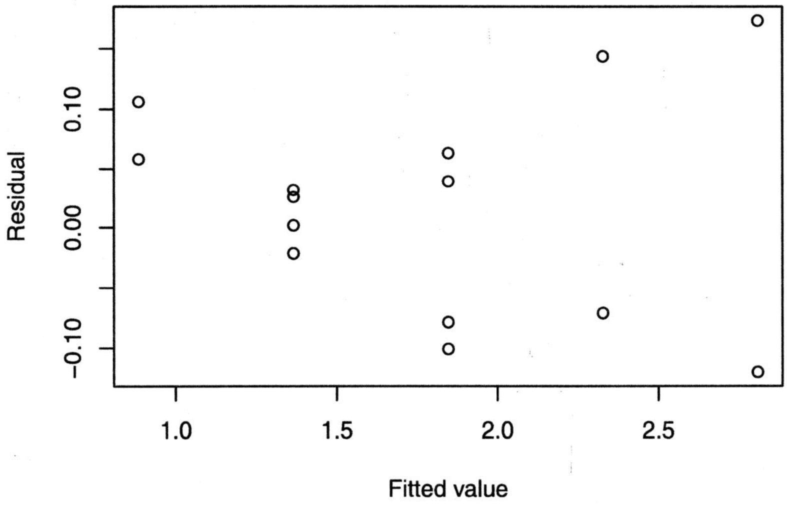

Chapter 12

ANALYSIS OF VARIANCE

12.1 (a) If the experiment is performed in soft water, the results may only be valid in soft water. Other kinds of water should also be used.

(b) With 15 results for detergent A and only 5 for detergent B, the variability for detergent A is known with much more precision than that for detergent B. Equal sample sizes should be used.

(c) The results may only apply to very hot water and very short wash times, and not to the circumstances of normal use with low temperature and longer washing times.

(d) There may be a time effect for the measurement process determining "whiteness". for example, the instrument may require recalibration after a few readings. The test results may be biased in this case.

12.3 (a) The grand mean is $\overline{y} = 578$. The deviations from the grand mean are:

Method A	Method B	Method C
-5	13	-6
-1	3	-1
-11	9	-2
-17	7	1

The sum of squares of the deviations is

$$SST = \sum_{i=1}^{k} \sum_{j=1}^{n_i} (y_{ij} - \overline{y}.)^2 = 546.$$

The means for the three samples are

Method A	Method B	Method C
572	586	576

The deviations of each sample from its own mean are

Method A	Method B	Method C
1	5	− 4
5	−5	1
−5	1	0
−1	− 1	3

The sum of squares of these deviations is

$$SSE = \sum_{i=1}^{k} \sum_{j=1}^{n_i} (y_{ij} - \overline{y}_j)^2 = 130.$$

The deviation of the individual sample means from the grand mean are

$$-6 \qquad 8 \qquad -2$$

Thus,

$$SS(Tr) = \sum_{i=1}^{k} n_i(\overline{y}_i - \overline{y}_.)^2 = 4(36) + 4(64) + 4(4) = 416.$$

(b) The total sample size $N = 4 + 4 + 4 = 12$, and the totals of each sample are

Method A	Method B	Method C
2288	2344	2304

Thus, $T_. = 2288 + 2344 + 2304 = 6,936$,

$$C = \frac{6936^2}{12} = 4,009,008 \quad \text{and} \quad \sum_{i=1}^{k} \frac{T_i^2}{n_i} = 4,009,424$$

The sum of squares of all of the observations is 73,554. Thus,

$$SST = \sum_{i=1}^{k} \sum_{j=1}^{n_i} y_{ij}^2 - C = 4,008,554 - 4,009,008 = 546.$$

$$SS(Tr) = \frac{\sum_{i=1}^{k} T_i^2}{n_i} - C = 4,009,424 - 4,009,008 = 416.$$

and

$$SSE = SST - SS(Tr) = 130.$$

These numbers agree with part (a).

12.5 The null hypothesis is that the mean number of mistakes is the same for the four technicians. The alternative is that the means are not the same. The analysis-of-variance table is

Source of variation	Degrees of freedom	Sum of squares	Mean square	F
Technicians	3	12.95	4.3167	0.68
Error	16	101.60	6.3500	
Total	19	114.55		

Since the critical value at the 0.01 level for an F distribution with 3 and 16 degrees of freedom is 5.29, we cannot reject the null hypothesis.

12.7 (a) We first find $k = 4$, $\bar{y} = 7$, $\bar{y}_1 = 5$, $\bar{y}_2 = 12$, $\bar{y}_3 = 9$, and $\bar{y}_4 = 3$. Thus,

$$\underset{y_{ij}}{\overset{\text{Obs.}}{\begin{bmatrix} 6 & 4 & 5 & & \\ 13 & 10 & 13 & 12 & \\ 7 & 9 & 11 & & \\ 3 & 6 & 1 & 4 & 1 \end{bmatrix}}} = \underset{\bar{y}}{\overset{\text{Grand mean}}{\begin{bmatrix} 7 & 7 & 7 & & \\ 7 & 7 & 7 & 7 & \\ 7 & 7 & 7 & & \\ 7 & 7 & 7 & 7 & 7 \end{bmatrix}}}$$

$$+ \underset{\bar{y}_i - \bar{y}}{\overset{\text{Tr. effect}}{\begin{bmatrix} -2 & -2 & -2 & & \\ 5 & 5 & 5 & 5 & \\ 2 & 2 & 2 & & \\ -4 & -4 & -4 & -4 & -4 \end{bmatrix}}} + \underset{y_{ij} - \bar{y}_i}{\overset{\text{Error}}{\begin{bmatrix} 1 & -1 & 0 & & \\ 1 & -2 & 1 & 0 & \\ -2 & 0 & 2 & & \\ 0 & 3 & -2 & 1 & -2 \end{bmatrix}}}$$

$$\begin{aligned} SS(Tr) &= 3(-2)^2 + 4(5)^2 + 3(2)^2 + 5(-4)^2 = 204, \quad df = 3. \\ SSE &= 1^2 + (-1)^2 + \cdots + 1^2 + (-2)^2 = 34, \quad df = 15 - 4 = 11. \\ SST &= 6^2 + 4^2 + \cdots + 4^2 + 1^2 - 15(7)^2 = 238, \quad df = 14 \\ &= SS(Tr) + SSE = 238. \ (check) \end{aligned}$$

(b) The null hypothesis is that the the four treatment population means are the same. The alternative is that they are not the same. The analysis-of-variance table is

Source of variation	Degrees of freedom	Sum of squares	Mean square	F
Treatments	3	204	68	22.00
Error	11	34	3.89	
Total	14	238		

Since the critical value at the 0.05 level for an F distribution with 3 and 11 degrees of freedom is 3.59, we can reject the null hypothesis. The treatment means are not the same.

12.9 The null hypothesis is that the true means of the samples of reactions times are the same. The alternative is that the true means are not the same. The analysis-of-variance table is

Source of variation	Degrees of freedom	Sum of squares	Mean square	F
Arrangements	2	81.429	40.7145	11.310
Error	25	90.000	3.600	
Total	27	171.429		

Since the critical value at the 0.01 level for an F distribution with 2 and 25 degrees of freedom is 5.57, we can reject the null hypothesis. The mean reaction times are not the same for the three arrangements.

12.11 The null hypothesis is that the true means of the samples of the six samples are the same. The alternative is that the true means are not the same. The analysis-of-variance table is

Source of variation	Degrees of freedom	Sum of squares	Mean square	F
Samples	5	1569.667	313.9334	15.697
Error	6	120.000	20.0000	
Total	11	1689.667		

Since the critical value at the 0.05 level for an F distribution with 5 and 6 degrees of freedom is 4.39, we reject the null hypothesis. These samples differ in mean compressive strength.

12.13 By assumption

$$P\left(|(\bar{Y_i} - \bar{Y_j}) - (\mu_i - \mu_j) > |E|\right) \geq \alpha\ , \qquad 1 \leq i \neq j \leq 3$$

Hence, for the equal tail test,

$$P\left(|Z| > E/\sigma\sqrt{\frac{1}{n_i} + \frac{1}{n_j}}\right) \geq \alpha$$

which implies

$$E/\sigma\sqrt{\frac{1}{n_i} + \frac{1}{n_j}} = z_{\alpha/2}$$

or

$$\frac{1}{n_i} + \frac{1}{n_j} = \left(\frac{E}{z_{\alpha/2}\sigma}\right)^2 = K \quad \text{where} \quad 1 \leq i \neq j \leq 3$$

That is, we require that

$$\frac{1}{n_1} = \frac{1}{n_2} = \frac{1}{n_3} = \frac{K}{2}$$

or

$$n_1 = n_2 = n_3 = \frac{2}{K} = 2\left(\frac{z_{\alpha/2}\sigma}{E}\right)^2 = 2\left(\frac{(1.96)(0.012)}{0.01}\right)^2 = 11.06$$

Therefore we take $n_1 = n_2 = n_3 = 12$.

12.15 By Theorem 12.1, using the notation $N = \sum_{i=1}^{k} n_i$,

$$\begin{aligned} SST &= \sum_{i=1}^{k}\sum_{j=1}^{n_i}(y_{ij} - \bar{y}_.)^2 = \sum_{i=1}^{k}\sum_{j=1}^{n_i}(y_{ij}^2 - 2y_{ij}\bar{y}_. + \bar{y}_.^2) \\ &= \sum_{i=1}^{k}\sum_{j=1}^{n_i} y_{ij}^2 - 2\bar{y}_.\sum_{i=1}^{k}\sum_{j=1}^{n_i} y_{ij} + N\bar{y}_.^2 \\ &= \sum_{i=1}^{k}\sum_{j=1}^{n_i} y_{ij}^2 - N\bar{y}_.^2 = \sum_{i=1}^{k}\sum_{j=1}^{n_i} y_{ij}^2 - C \end{aligned}$$

since $N\bar{y}_.^2 = T_.^2/N$.

Similarly,

$$\begin{aligned} SS(Tr) &= \sum_{i=1}^{k} n_i(\bar{y}_i - \bar{y}_.)^2 = \sum_{i=1}^{k} n_i\bar{y}_i^2 - 2\bar{y}_.\sum_{i=1}^{k} n_i\bar{y}_i + N\bar{y}_.^2 \\ &= \sum_{i=1}^{k} n_i\left(\frac{T_i}{n_i}\right)^2 - 2\frac{T_.}{N}\sum_{i=1}^{k} n_i\frac{T_i}{n_i} + N\left(\frac{T_.}{N}\right)^2 \\ &= \sum_{i=1}^{k}\frac{T_i^2}{n_i} - 2\frac{T_.}{N}\sum_{i=1}^{k} T_i + \frac{T_.^2}{N} \\ &= \sum_{i=1}^{k}\frac{T_i^2}{n_i} - \frac{T_.^2}{N} = \sum_{i=1}^{k}\frac{T_i^2}{n_i} - C \end{aligned}$$

12.17 (a) The null hypothesis is that the true mean aflatoxin content is the same for both brands. The alternative is that they are different. The analysis-of-variance table is

Source of variation	Degrees of freedom	Sum of squares	Mean square	F
Brands	1	11.734	11.7340	1.047
Error	12	134.515	11.2096	
Total	13	146.249		

Since the critical value at the 0.05 level for an F distribution with 1 and 12 degrees of freedom is 4.75, we cannot reject the null hypothesis. The two brands do not differ in mean aflatoxin content.

(b) The two-sample t statistic for testing the null hypothesis $\delta = 0$ versus the alternative hypothesis

$\delta \neq 0$, is given by

$$t = \frac{\bar{y}_1 - \bar{y}_2}{\sqrt{(n_1-1)s_1^2 + (n_2-1)s_2^2}} \sqrt{\frac{n_1 n_2 (n_1 + n_2 - 2)}{n_1 + n_2}}$$

In this case $\bar{y}_1 = 2.2$, $\bar{y}_2 = 4.05$, $s_1^2 = 8.157$ and $s_2^2 = 15.483$. Since $n_1 = 8$ and $n_2 = 6$,

$$t = \frac{2.2 - 4.05}{\sqrt{(8-1)8.157 + (6-1)15.483}} \sqrt{\frac{(8)(6)(8+6-2)}{8+6}} = -1.023$$

Since the .025 value of a t with 12 degrees of freedom is 2.179. we cannot reject the null hypothesis.

(c) Notice that, within round off, $t^2 = (-1.023)^2 = 1.047 = F$ for the statistics and that $t_\nu^2\,(2.179)^2 = 4.75 = F(1, \nu)$ for the critical values. The two analyzes are equivalent, since both statistics will reject and accept at the same time.

12.19 (a) There are $a = 3$ treatments so that the degrees of freedom for Total are $ab - 1 = 11$ so $b = 12/3 = 4$. The Block sum of squares has degrees of freedom $= b - 1 = 4 - 1 = 3$. Next,

$$\bar{y}_{..} = \frac{4\bar{y}_{1\cdot} + 4\bar{y}_{2\cdot} + 4\bar{y}_{3\cdot}}{3 \times 4} = \frac{96}{12} = 8$$

so the sum of squares treatment

$$SS(Tr) = b \sum_{i=1}^{a} (\bar{y}_{i\cdot} - \bar{y}_{..}) = 4(-2)^2 + 4(-1)^2 + 4(3)^2 = 56$$

The treatment sum of squares has df $= a - 1 = 3 - 1 = 2$.

Finally, by subtraction, the error sum of squares

$$SSE = SST - SS(Tr) - SS(Bl) = 220 - 56 - 132 = 32$$

with $(a-1)(b-1) = 2 \cdot 3 = 6$ degrees of freedom

The completed analysis-of-variance table is

Source of variation	Degrees of freedom	Sum of squares	Mean square	F
Treatment	2	56	28	5.250
Blocks	3	132	44	8.251
Error	6	32	5.333	
Total	11	220		

(b) Since the critical value at the 0.05 level for an F distribution with 2 and 6 degrees of freedom is 5.14, we reject the null hypothesis of no differences in the treatment means. Further, since the critical value at the 0.05 level for an F distribution with 3 and 6 degrees of freedom is 4.76, we reject the null hypothesis of no differences in blocks.

12.21 The null hypothesis is that the true means for the technicians are the same. The alternative hypothesis is that they are not the same. A second null hypothesis is that there is no block(day) effect.

The analysis-of-variance table is

Source of variation	Degrees of freedom	Sum of squares	Mean square	F
Technicians	3	12.95	4.32	0.831
Days	4	20.80	5.20	0.773
Error	12	80.8	6.73	
Total	19	114.55		

Since the critical value at the 0.01 level for an F distribution with 3 and 12 degrees of freedom is 5.95, we cannot reject the null hypothesis that there is no difference between the means for the technicians. Since the critical value at the 0.01 level for an F distribution with 4 and 12 degrees of freedom is 5.41, we cannot reject the null hypothesis of no block effect.

12.23 (a) We first find $a = 4$, $\bar{y}_{..} = 10$, $\bar{y}_{1.} = 8$, $\bar{y}_{2.} = 13$, $\bar{y}_{3.} = 10$ and $\bar{y}_{4.} = 9$. And $b = 5$, $\bar{y}_{.1} = 13$, $\bar{y}_{.2} = 8$, $\bar{y}_{.3} = 12$, $\bar{y}_{.4} = 6$ and $\bar{y}_{.5} = 11$. Thus,

$$\underset{\text{Obs.}}{\underset{y_{ij}}{\begin{bmatrix} 14 & 6 & 11 & 0 & 9 \\ 14 & 10 & 16 & 9 & 16 \\ 12 & 7 & 10 & 9 & 12 \\ 12 & 9 & 11 & 6 & 7 \end{bmatrix}}} = \underset{\text{Grand mean}}{\underset{\bar{y}_{..}}{\begin{bmatrix} 10 & 10 & 10 & 10 & 10 \\ 10 & 10 & 10 & 10 & 10 \\ 10 & 10 & 10 & 10 & 10 \\ 10 & 10 & 10 & 10 & 10 \end{bmatrix}}} + \underset{\text{Tr. effect}}{\underset{\bar{y}_{i.} - \bar{y}_{..}}{\begin{bmatrix} -2 & -2 & -2 & -2 & -2 \\ 3 & 3 & 3 & 3 & 3 \\ 0 & 0 & 0 & 0 & 0 \\ -1 & -1 & -1 & -1 & -1 \end{bmatrix}}}$$

$$+ \underset{\text{Bl. effect}}{\underset{\bar{y}_{.j} - \bar{y}_{..}}{\begin{bmatrix} 3 & -2 & 2 & -4 & 1 \\ 3 & -2 & 2 & -4 & 1 \\ 3 & -2 & 2 & -4 & 1 \\ 3 & -2 & 2 & -4 & 1 \end{bmatrix}}} + \underset{\text{Error}}{\underset{y_{ij} - \bar{y}_{i.} - \bar{y}_{.j} + \bar{y}_{..}}{\begin{bmatrix} 3 & 0 & 1 & -4 & 0 \\ -2 & -1 & 1 & 0 & 2 \\ -1 & -1 & -2 & 3 & 1 \\ 0 & 2 & 0 & 1 & -3 \end{bmatrix}}}$$

(b) The sums of squares and degrees of freedoms are

$$\begin{aligned} SS(Tr) &= 5(-2)^2 + 5(3)^2 + 5(0)^2 + 5(-1)^2 = 70, \quad df = 3. \\ SS(Bl) &= 4(3)^2 + 4(-2)^2 + 4(2)^2 + 4(-4)^2 + 4(1) = 136, \quad df = 4. \end{aligned}$$

$$SSE = 3^2 + 0^2 + 1^2 + \cdots + 0^2 + 1^2 + (-3)^2 = 66,$$

$$df = (4-1)(5-1) = 12.$$

$$SST = 14^2 + 6^2 + 11^2 + \cdots + 11^2 + 6^2 + 7^2 - 20(10)^2 = 272,$$

$$df = 19$$

$$= SS(Tr) + SS(Bl) + SSE = 272. \quad (check)$$

(c) The null hypothesis is that the the four treatment population means are the same. The alternative is that they are not the same. The analysis-of-variance table is

Source of variation	Degrees of freedom	Sum of squares	Mean square	F
Treatments	3	70	23.33	4.24
Blocks	4	136	34.00	6.18
Error	12	66	5.50	
Total	19	272		

Since the critical value at the 0.05 level for an F distribution with 3 and 12 degrees of freedom is 3.49, we can reject the null hypothesis. The treatment means are not the same. Since $F_{.05}$ with 4 and 12 degrees of freedom is 3.26, the block effect of the experiment is also apparent.

12.25 The null hypothesis is that the true means for the detergents are the same. The alternative hypothesis is that they are not the same.

The one-way analysis-of-variance table is

Source of variation	Degrees of freedom	Sum of squares	Mean square	F
Detergents	3	110.92	36.97	1.95
Error	8	154.00	19.25	
Total	11	264.92		

Since the critical value at the 0.05 level for an F distribution with 3 and 8 degrees of freedom is 4.07, we cannot reject the null hypothesis of no detergent effect. In the two-way analysis, we could reject the null hypothesis at the $\alpha = 0.01$ level. This shows how important it is to be sure that there is no systematic variation in the error sum of squares.

12.27 To use the sums of squares formula, we need to compute the marginal and total sums

$$T_{..1} = 438, \quad T_{..2} = 483$$

$$T_{.1.} = 223, \quad T_{.2.} = 236, \quad T_{.3.} = 242, \quad T_{.4.} = 220$$

$$T_{1..} = 308, \quad T_{2..} = 304, \quad T_{3..} = 309, \quad T_{...} = 921$$

Thus, $C = (921)^2/(3 \cdot 4 \cdot 2) = 35,343.375$. Further,

$$\sum_{i=1}^{a} T_{i..}^2 = 308^2 + 304^2 + 309^2 = 282,761$$

$$\sum_{j=1}^{b} T_{.j.}^2 = 223^2 + 236^2 + 242^2 + 220^2 = 212,389$$

$$\sum_{k=1}^{r} T_{..k}^2 = 438^2 + 483^2 = 425,133$$

$$\sum_{i=1}^{a}\sum_{j=1}^{b}\sum_{k=1}^{r} y_{ijk}^2 = 35,715$$

Thus,

$$SST = 35,715 - 35,343.375 = 371.625$$

$$SS(Tr) = \frac{282,761}{4 \cdot 2} - 35,343.375 = 1.750$$

$$SS(Bl) = \frac{212,389}{3 \cdot 2} - 35,343.375 = 54.792$$

$$SS(Reps) = \frac{425,133}{4 \cdot 3} - 35,343.375 = 84.375$$

$$SSE = 371.62 - 1.745 - 54.79 - 84.37 = 230.715$$

The analysis of variance is

Source of variation	Degrees of freedom	Sum of squares	Mean square	F
Machines	2	1.750	0.875	0.064
Workers	3	54.792	18.264	1.364
Reps	1	84.375	84.375	6.217
Error	17	230.708	13.571	
Total	23	371.625		

Since the critical value at the 0.05 level for an F distribution with 2 and 17 degrees of freedom is 3.59, we cannot reject the null hypothesis of no treatment (machine) effect.

Since the critical value at the 0.05 level for an F distribution with 3 and 17 degrees of freedom is 3.20, we cannot reject the null hypothesis of no block(worker) effect.

Since the critical value at the 0.05 level for an F distribution with 1 and 17 degrees of freedom is 4.45, we reject the null hypothesis of no replication effect.

12.29 We are given that

$$\mu_{ij} = \mu + \alpha_i + \beta_j, \qquad \frac{1}{b}\sum_{j=1}^{b} \mu_{ij} = \mu + \alpha_i, \qquad \frac{1}{ab}\sum_{i=1}^{a}\sum_{j=1}^{b} \mu_{ij} = \mu$$

Summing the first relation over j and dividing by b gives

$$\frac{1}{b}\sum_{j=1}^{b} \mu_{ij} = \mu + \alpha_i + \frac{1}{b}\sum_{j=1}^{b} \beta_j$$

Comparing this with the second given relation, we conclude that

$$\frac{1}{b}\sum_{j=1}^{b} \beta_j = 0 \quad or \quad \sum_{j=1}^{b} \beta_j = 0$$

Next, summing the first given relation over i and j and dividing by ab,

$$\frac{1}{ab}\sum_{i=1}^{a}\sum_{j=1}^{b} \mu_{ij} = \mu + \frac{1}{a}\sum_{i=1}^{a} \alpha_i + \frac{1}{b}\sum_{j=1}^{b} \beta_j$$

Comparing this with the third given relation, and using $\sum_{j=1}^{b} \beta_j = 0$, we conclude that $\sum_{k=1}^{a} \alpha_i = 0$,

12.31 From Exercise 12.21, the (sorted) means for the four treatments are

Treatment 1	Treatment 4	Treatment 3	Treatment 2
8	9	10	13

and the $MSE = 5.5$ with 5 degrees of freedom. From Table 8(a), after multiplying each r_p by $s_{\bar{y}} = \sqrt{MSE/n} = \sqrt{5.5/5} = 1.049$ to get R_p, we obtain

p	2	3	4
r_p	3.08	3.23	3.31
R_p	3.23	3.39	3.47

For $\alpha = .05$, none of the ranges between two adjacent means are significant. Only $\bar{y}_2 - \bar{y}_4 = 4$ is significant among the ranges of three adjacent means. The range of four means $\bar{y}_2 - \bar{y}_1 = 5$ is significant among the ranges of three adjacent means. The range of four means $\bar{y}_2 - \bar{y}_1 = 5 > R_4$ is also significant. We conclude

Tr 1	Tr 4	Tr 3	Tr 2
8	9	10	13

12.33 The (sorted) means for the three methods are

A	C	B
572	576	586

and the $MSE = 130/9 = 14.44$ with 9 degrees of freedom. and $s_{\bar{y}} = \sqrt{MSE/n} = \sqrt{14.44/4} = 1.9$ With $k = 3$, $\alpha = .03$ and $\alpha/k(k-1) = .03/6 = .005$, we have $t_{.005} = 3.250$ with 9 degrees of freedom. Using Bonferroni's procedure, we can construct 3 confidence intervals for the yields under the three settings of the magnetic field and we have 97% confidence that all 3 intervals hold simultaneously. For example, the confidence interval for $\mu_1 - \mu_3$ is

$$\begin{aligned} \mu_1 - \mu_3 \ : \quad & \bar{y}_{1.} - \bar{y}_{3.} \pm t_{.005}\sqrt{MSE(\frac{1}{4} + \frac{1}{4})} \\ &= (572 - 576) \pm 3.250\sqrt{14.44\,(\frac{1}{4} + \frac{1}{4})} \\ &= -4 \pm 8.73 \end{aligned}$$

Similarly, with the same 8.73 applying for all pairs of mean differences, we obtain

$$\begin{aligned} \mu_1 - \mu_2 \ : \quad & (572 - 586) \pm 8.73 = -14 \pm 8.73 \\ \mu_2 - \mu_3 \ : \quad & (586 - 576) \pm 8.73 = 10 \pm 8.73 \end{aligned}$$

With 97 % confidence, the second setting has higher mean yields It has from 5.37 to 22.73 higher than the first setting and 1.37 to 18.73 higher than the third.

12.35 (a) The (sorted) sample means for the five threads are

thread 1	thread 5	thread 3	thread 4	thread 2
20.675	20.900	23.525	23.700	25.650

and, from Exercise 12.20, the $MSE = 2.110$ with 12 degrees of freedom. From Table 8(b) with 12 degrees of freedom , after multiplying each r_p by $s_{\bar{y}} = \sqrt{MSE/n} = \sqrt{2.110/4} = 0.726$ to get R_p, we obtain

p	2	3	4	5
r_p	4.32	4.50	4.62	4.71
R_p	3.14	3.27	3.35	3.42

For $\alpha = .01$, none of the ranges between two adjacent means, or three adjacent means, are significant. Only $\bar{y}_2 - \bar{y}_5 = 4.750 > R_4$ is significant among the ranges of four adjacent means. The range of all five means $\bar{y}_2 - \bar{y}_1 = 4.975 > R_5$ is also significant.

thread 1	thread 5	thread 3	thread 4	thread 2
20.675	20.900	23.525	23.700	25.650

(b) With $k = 5$, $\alpha = .10$ and $\alpha/k(k-1) = .10/20 = .005$, we have $t_{.005} = 3.055$ with 12 degrees of freedom. Using Bonferroni's procedure, we can construct 10 confidence intervals for the differences of mean breaking strength of the five threads and are 90% confidence that all 10 intervals hold simultaneously. For example, the confidence interval for $\mu_1 - \mu_2$ is

$$\begin{aligned}\mu_1 - \mu_2 \ : \quad & (\bar{y}_{1.} - \bar{y}_{2.}) \pm t_{.005}\sqrt{MSE(\frac{1}{4} + \frac{1}{4})} \\ &= (20.675 - 25.650) \pm 3.055\sqrt{2.11(\frac{1}{4} + \frac{1}{4})} \\ &= -4.975 \pm 3.138\end{aligned}$$

Similarly, with the same 3.138 applying for all pairs of mean differences, we obtain

$$\begin{aligned}\mu_1 - \mu_3 &: (20.675 - 23.525) \pm 3.138 = -2.850 \pm 3.138 \\ \mu_1 - \mu_4 &: (20.675 - 23.700) \pm 3.138 = -3.025 \pm 3.138 \\ \mu_1 - \mu_5 &: (20.675 - 20.900) \pm 3.138 = -0.225 \pm 3.138 \\ \mu_2 - \mu_3 &: (25.650 - 23.525) \pm 3.138 = 2.125 \pm 3.138 \\ \mu_2 - \mu_4 &: (25.650 - 23.700) \pm 3.138 = 1.950 \pm 3.138 \\ \mu_2 - \mu_5 &: (25.650 - 20.900) \pm 3.138 = 4.750 \pm 3.138 \\ \mu_3 - \mu_4 &: (23.525 - 23.700) \pm 3.138 = -0.175 \pm 3.138 \\ \mu_3 - \mu_5 &: (23.525 - 20.900) \pm 3.138 = 2.625 \pm 3.138 \\ \mu_4 - \mu_5 &: (23.700 - 20.900) \pm 3.138 = 2.800 \pm 3.138\end{aligned}$$

Note that only two confidence intervals are not include zero. We are 90% confident that in average, thread 2 is stronger than threads 1 and 5, but also are unable to differentiate between the mean strength of threads 1, 3, 4, and 5 or threads 2, 3 and 4.

12.37 The null hypothesis is that there is no treatment effect, that is there is no difference in the effects of the rebuilding methods on the mean times to failure. The level is $\alpha = .05$. The analysis-of-covariance table is

Source of variation	Sum of squares for x	Sum of squares for y	Sum of products	Sum of Squares for y'	Degrees of freedom	Mean square
Treatments	26	6	−3	10.759	2	5.3795
Error	28	10	14	3	5	0.6
Total	54	16	11	13.759	7	

Thus, the F statistic is 5.3795/0.6 =8.966. Since the critical value at the 0.05 level for an F distribution with 2 and 5 degrees of freedom is 5.79, we reject the null hypothesis at level $\alpha = 0.05$.

The regression coefficient is given by $SPE/SSE_x = 14/28 = 0.5$.

12.39 The null hypothesis is that track designs are equally resistant to breakage. The level is $\alpha = .01$. The analysis-of-covariance table is

Source of variation	Sum of squares for x	Sum of squares for y	Sum of products	Sum of Squares for y'	Degrees of freedom	Mean square
Treatments	420.130	114.55	192.33	26.723	3	8.908
Error	556.052	122.40	237.74	20.754	15	1.384
Total	976.182	236.95	430.07	47.477	18	

Thus, the F statistic is 6.44. Since the critical value at the 0.01 level for an F distribution with 3 and 15 degrees of freedom is 5.42, we reject the null hypothesis.

The effect of usage on breakage resistance, that is, the regression coefficient of the model, is given by $SPE/SSE_x = .428$.

12.41 The $P-$value for the slope of leakage current suggests that it is not necessary to adjust for this covariate.

```
Analysis of Variance for y, using Adjusted SS for Tests


Source      DF    Seq SS    Adj SS   Adj MS       F      P
x            1     56019    182056   182056    3.64  0.067
Condition    2   1027473   1027473   513736   10.28  0.001
Error       26   1299902   1299902    49996
Total       29   2383393



S = 223.598   R-Sq = 45.46%   R-Sq(adj) = 39.17%
```

```
Term          Coef   SE Coef      T      P
Constant   10023.1     202.6  49.46  0.000
x           -57.02     29.88  -1.91  0.067
```

12.43 The null hypothesis is that the weight losses for the three lubricants are the same. The alternative is that the weight losses are not the same. The analysis-of-variance table is

Source of variation	Degrees of freedom	Sum of squares	Mean square	F
Flows	3	0.528	0.1760	2.80
Error	16	1.004	0.0628	
Total	19	1.532		

Since the critical value at the 0.05 level for an F distribution with 3 and 16 degrees of freedom is 3.24, we cannot reject the null hypothesis that flow through the precipitator has no effect on the exit loading.

12.45 (a) We first find $a = 3$, $\bar{y}_{..} = 9$, $\bar{y}_{1.} = 7$, $\bar{y}_{2.} = 8$ and $\bar{y}_{3.} = 12$. And $b = 4$, $\bar{y}_{.1} = 8$, $\bar{y}_{.2} = 13$, $\bar{y}_{.3} = 4$ and $\bar{y}_{.4} = 11$. Thus,

$$\underset{y_{ij}}{\overset{\text{Obs.}}{\begin{bmatrix} 9 & 10 & 2 & 7 \\ 6 & 3 & 1 & 12 \\ 9 & 16 & 9 & 14 \end{bmatrix}}} = \underset{\bar{y}_{..}}{\overset{\text{Grand mean}}{\begin{bmatrix} 9 & 9 & 9 & 9 \\ 9 & 9 & 9 & 9 \\ 9 & 9 & 9 & 9 \end{bmatrix}}} + \underset{\bar{y}_{i.} - \bar{y}_{..}}{\overset{\text{Tr. effect}}{\begin{bmatrix} -2 & -2 & -2 & -2 \\ -1 & -1 & -1 & -1 \\ 3 & 3 & 3 & 3 \end{bmatrix}}}$$

$$+ \underset{\bar{y}_{.j} - \bar{y}_{..}}{\overset{\text{Bl. effect}}{\begin{bmatrix} -1 & 4 & -5 & 2 \\ -1 & 4 & -5 & 2 \\ -1 & 4 & -5 & 2 \end{bmatrix}}} + \underset{y_{ij} - \bar{y}_{i.} - \bar{y}_{.j} + \bar{y}_{..}}{\overset{\text{Error}}{\begin{bmatrix} 3 & -1 & 0 & -2 \\ -1 & 1 & -2 & 2 \\ -2 & 0 & 2 & 0 \end{bmatrix}}}$$

(b) The sums of squares and degrees of freedoms are

$$\begin{aligned} SS(Tr) &= 4(-2)^2 + 4(-1)^2 + 4(3)^2 = 56, \quad df = 2 \\ SS(Bl) &= 3(-1)^2 + 3(4)^2 + 3(-5)^2 + 3(2)^2 = 138, \quad df = 3 \\ SSE &= 3^2 + (-1)^2 + 0^2 + \cdots + 0^2 + 2^2 + 0^2 = 32, \\ & \quad df = (3-1)(4-1) = 6 \end{aligned}$$

$$SST = 9^2 + 10^2 + 2^2 + \cdots + 16^2 + 9^2 + 14^2 - 12(9)^2 = 226,$$
$$df = 11$$
$$= SS(Tr) + SS(Bl) + SSE = 226. \quad (check)$$

(c) The null hypothesis is that the the four treatment population means are the same. The alternative is that they are not the same. A second null hypothesis is that there is no block effect. The analysis-of-variance table is

Source of variation	Degrees of freedom	Sum of squares	Mean square	F
Treatments	2	56	28.00	5.25
Blocks	3	138	46.00	8.63
Error	6	32	5.33	
Total	11	226		

Since the critical value at the 0.05 level for an F distribution with 2 and 6 degrees of freedom is 5.14, we can reject the null hypothesis. The treatment means are not the same. Since $F_{.05}$ with 3 and 6 degrees of freedom is 4.76, the block effect of the experiment is also apparent.

12.47 The null hypothesis is that there is no agency effect. A second null hypothesis is that there is no site effect. The analysis-of-variance table is

Source of variation	Degrees of freedom	Sum of squares	Mean square	F
Agency	2	26.572	13.286	4.84
Sites	4	1,117.263	279.316	101.75
Error	8	21.961	2.745	
Total	14	1,165.746		

(a) Since the critical value at the 0.05 level for an F distribution with 2 and 8 degrees of freedom is 4.46, we reject the null hypothesis that there is no agency effect.

(b) Since the critical value at the 0.05 level for an F distribution with 4 and 8 degrees of freedom is 3.84, we also reject the null hypothesis of no site effect. We conclude that the agencies are not consistent and that the sites have different levels of contamination.

12.49 (a) The grand mean is $\overline{y} = 4$. The deviations from the grand mean are:

Method A	Method B	Method C
−2	0	1
1	2	−1
−2	1	0

The sum of squares of the deviations is

$$SST = \sum_{i=1}^{k} \sum_{j=1}^{n_i} (y_{ij} - \overline{y}_{\cdot})^2 = 16.$$

The means for the three samples are

Method A	Method B	Method C
3	5	4

The deviations of each sample from its own mean are

Method A	Method B	Method C
−1	−1	1
2	1	−1
−1	0	0

The sum of squares of these deviations is

$$SSE = \sum_{i=1}^{k} \sum_{j=1}^{n_i} (y_{ij} - \overline{y_j})^2 = 10.$$

The deviation of the individual sample means from the grand mean are

$$-1 \quad 1 \quad 0$$

Thus,

$$SS(Tr) = \sum_{i=1}^{k} n_i(\overline{y_i} - \overline{y}_{\cdot})^2 = 3(1) + 3(1) + 3(0) = 6.$$

The analysis-of-variance table is

Source of variation	Degrees of freedom	Sum of squares	Mean square	F
Methods	2	6	3.0000	1.80
Error	6	10	1.6667	
Total	8	16		

Since the critical value at the 0.05 level for an F with (2,6) degrees of freedom is 5.14, we fail to reject the null hypothesis of equal mean time to failure under the three methods.

(b) When life length is adjusted for the age of robotic welder, we do establish a difference in mean time to failure.

12.51 (a) The null hypothesis is that there is no difference in the mean roughness under the three surface treatments. The level is $\alpha = .05$. The analysis-of-covariance table is

Source of variation	Sum of squares for x	Sum of squares for y	Sum of products	Sum of Squares for y'	Degrees of freedom	Mean square
Treatments	6	78	21	15.3375	2	7.6688
Error	10	36	18	3.6000	5	0.7200
Total	16	114	39	18.9375	7	

Thus, the F statistic is $7.6688/0.72 = 10.652$. Since the critical value at the 0.05 level for an F distribution with 2 and 6 degrees of freedom is 5.14, we reject the null hypothesis.

The regression coefficient is given by $SPE/SSE_x = 18/10 = 1.8$.

(b) The value $F = 10.652$ is much larger here although there is one less degree of freedom and so this F is more variable than the one in Exercise 12.50. The $P-$ value would be smaller here.

12.53 To verify the analysis-of-variance table, we need to find the treatment sums of the sample.

$$T_{1.} = 7.25, \quad T_{2.} = 3.35, \quad T_{3.} = 7.10 \quad T_{4.} = 7.23.$$

Hence

$$T_{..} = 24.93 \quad \text{and} \quad C = \frac{24.93^2}{(4)(3)} = 51.7921.$$

Since $k = 4$, $n = 3$ and

$$\sum_{i=1}^{k} \sum_{j=1}^{n} y_{ij}^2 = 73.8769,$$

we have

$$SST = \sum_{i=1}^{k} \sum_{j=1}^{n} y_{ij}^2 - C^2 = 73.8769 - 51.7921 = 22.0848.$$

Now

$$\sum_{i=1}^{k} \frac{T_i^2}{n} = \frac{7.25^2}{3} + \frac{3.35^2}{3} + \frac{7.10^2}{3} + \frac{7.23^2}{3} = 55.4893.$$

Hence

$$SS(Tr) = \sum_{i=1}^{k} \frac{T_i^2}{n} - C = 55.4893 - 51.7921 = 3.6972$$

and

$$SSE = SST - SS(Tr) = 22.0848 - 3.6972 = 18.3876.$$

The F statistic is

$$F = \frac{MS(Tr)}{MSE} = \frac{3.6972/3}{18.3876/8} = .54.$$

Hence the analysis-of-variance table is

Source of variation	Degrees of freedom	Sum of squares	Mean square	F
Paper type	3	3.70	1.23	.54
Error	8	18.39	2.30	
Total	11	22.08		

which is the same as that shown in the text.

Chapter 13

FACTORIAL EXPERIMENTATION

13.1 1. We test each of the following hypotheses.

Null hypothesis (a): The replication effect is zero. $\rho_1 = \rho_2 = \rho_3 = 0$.

We reject the null hypothesis at .05 level if $F > F_{.05} = 4.84$ with 1 and 11 degrees of freedom.

Null hypothesis (b): The two-factor interaction is zero.

We reject the null hypothesis at .05 level if $F > F_{.05} = 2.93$ with 4 and 18 degrees of freedom.

Null hypothesis (c): The temperature(factor B) has no effect. $\beta_1 = \beta_2 = \beta_3 = \beta_4 = \beta_5 = 0$.

We reject the null hypothesis at .05 level if $F > F_{.05} = 2.93$ with 4 and 18 degrees of freedom.

Null hypothesis (d): The concentration(factor A) has no effect. $\alpha_1 = \alpha_2 = 0$.

We reject the null hypothesis at .05 level if $F > F_{.05} = 4.41$ with 1 and 18 degrees of freedom.

2. **Calculations:** The data are

Concentration grams/liter	Temperature (degrees F)	Reflectivity Rep 1	Rep 2	Rep 3	Sum
5	75	35	39	36	110
5	100	31	37	36	104
5	125	30	31	33	94
5	150	28	20	23	71
5	175	19	18	22	59
10	75	38	46	41	125
10	100	36	44	39	119
10	125	39	32	38	109
10	150	35	47	40	122
10	175	30	38	31	99
	Total	321	352	339	1,012

And the table of totals for the two factors are

		Factor B: Temperature(F^0)					
		75	100	125	150	175	
Factor A:	5	110	104	94	71	59	438
Concentration	10	125	119	109	122	99	574
		235	223	203	193	158	1,012

Hence

$$C = \frac{1,012^2}{30} = 34,138.1333$$

And the sum of squares are

$$\begin{aligned}
SST &= \sum_i \sum_j \sum_k y_{ijk}^2 - C = 35,822 - C = 1,683.8667 \\
SS(Tr) &= \frac{1}{r} \sum_i \sum_j T_{ij.}^2 - C = \frac{1}{3}(110^2 + 104^2 + \cdots + 122^2 + 99^2) - C \\
&= 1,403.8667 \\
SSR &= \frac{1}{ab} \sum_k T_{..k}^2 - C = \frac{1}{(2)(5)}(321^2 + 352^2 + 339^2) - C = 48.4667 \\
SSE &= SST - SS(Tr) - SSR = 231.5333 \\
SSA &= \frac{1}{br} \sum_{i=1}^{a} T_{i..}^2 - C = \frac{1}{(5)(3)}(438^2 + 574^2) - C = 616.5334 \\
SSB &= \frac{1}{ar} \sum_{j=1}^{b} T_{i.j.}^2 - C \\
&= \frac{1}{(2)(3)}(235^2 + 223^2 + 203^2 + 193^2 + 158^2) - C = 591.2000 \\
SS(AB) &= SS(Tr) - SSA - SSB = 196.1333
\end{aligned}$$

The analysis-of-variance table is

Source of variation	Degrees of freedom	Sums of squares	Mean square	F
Replication	2	48.4667	24.2334	1.88
A Concentration	1	616.5334	616.5334	47.93
B Temperature	4	591.2000	147.8000	11.49
AB interaction:	4	196.1333	49.0333	3.81
Error	18	231.5333	12.8630	
Total	29	1683.8667		

3. **Decision:** The replication effect is not significant at .05 level. The concentration effect and the temperature effect are both significant at .05 level. However, these cannot be interpreted individually because the interaction effect is significant at .05 level.

   The two-way table of fitted values, $\bar{y}_{ij.}$, provides a summary of the experiment.

		Factor B: Temperature(F^0)				
		75	100	125	150	175
Factor A:	5	36.67	34.67	31.33	23.67	19.67
Concentration	10	41.67	39.67	36.33	40.67	33.00

4. **Further Analysis:** The optimal reflectivity occurred at 75^0F with concentration 10 grams per liter. At optimal conditions, the reflectivity is a normal random variable with mean μ_{21} and variance σ^2. Since the mean square error(MSE) is an unbiased estimate for σ^2 and $t_{.025} = 2.101$ with 18 degrees of freedom, the 95% confidence interval of reflectivity at optimal conditions is

$$\bar{y}_{21.} \pm t_{.025}\sqrt{\frac{MSE}{n}} = 41.67 \pm 2.101\sqrt{\frac{12.8630}{3}} = 41.67 \pm 4.35$$

or $37.32 < \mu_{21} < 46.02$.

13.3 The table of totals for the two factors are

		Engine 1	Engine 2	Engine 3	
Detergent	A	84	85	109	278
	B	91	92	100	283
	C	82	97	100	279
	D	89	82	106	277
		346	356	415	1117

We also have $a = 4$, $b = 3$, $r = 2$ and

$$T_{..1} = 565, \quad T_{..2} = 552, \quad \sum_{i=1}^{4}\sum_{j=1}^{3}\sum_{k=1}^{2} y_{ijk}^2 = 52,745$$

Hence $C = 1117^2/(4)(3)(2) = 51,987.04$.

$$\begin{aligned}
SST &= \sum_i\sum_j\sum_k y_{ijk}^2 - C = 52,745 - 51,987.04 = 757.96\\
SS(Tr) &= \frac{1}{r}\sum_i\sum_j T_{ij.}^2 - C = \frac{1}{2}(84^2 + 85^2 + \cdots + 82^2 + 106^2) - 51,987.04\\
&= 473.46\\
SSR &= \frac{1}{ab}\sum_k T_{..k}^2 - C = \frac{1}{(4)(3)}(565^2 + 552^2) - 51,987.04 = 7.04\\
SSE &= SST - SS(Tr) - SSR = 277.46\\
SSA &= \frac{1}{br}\sum_{i=1}^{a} T_{i..}^2 - C = \frac{1}{(3)(2)}(278^2 + 283^2 + 279^2 + 277^2) - 51,987.04\\
&= 3.46
\end{aligned}$$

$$\begin{aligned} SSB &= \frac{1}{ar}\sum_{j=1}^{b} T_{.j.}^2 - C = \frac{1}{(4)(2)}(346^2 + 356^2 + 415^2) - 51,987.04 \\ &= 347.59 \\ SS(AB) &= SS(Tr) - SSA - SSB = 122.41 \end{aligned}$$

The analysis-of-variance table is

Source of variation	Degrees of freedom	Sums of squares	Mean square	F
Replication	1	7.04	7.04	0.28
A (Detergent)	3	3.46	1.15	.05
B (Engine)	2	347.59	173.80	6.89
AB interaction:	6	122.41	20.40	0.81
Error	11	277.46	25.22	
Total	23	757.96		

Since $F_{.05} = 4.84$ with 1 and 11 degrees of freedom, the replication effect is not significant at the .05 level. Since $F_{.05} = 3.59$ with 3 and 11 degrees of freedom, the detergent effect is not significant at .05 level. Since $F_{.05} = 3.98$ with 2 and 11 degrees of freedom, the engine effect is significant at .05 level. Since $F_{.05} = 3.09$ with 6 and 11 degrees of freedom, the interaction effect is not significant at .05 level.

We conclude that the engine has an important effect. Machine 3 has a higher "cleanness" level than the other engines. All other effects may be unimportant.

13.5 The table of totals for the two factors are

		Filler 32	Filler 37	Filler 42	
Flow	3	4.09	4.64	5.17	13.90
	12	4.63	4.61	5.13	14.37
	30	4.65	5.11	5.20	14.96
		13.37	14.36	15.50	43.23

We also have $a = 3$, $b = 3$, $r = 2$ and

$$T_{..1} = 21.75, \quad T_{..2} = 21.48, \quad \sum_{i=1}^{3}\sum_{j=1}^{3}\sum_{k=1}^{2} y_{ijk}^2 = 104.4311$$

Hence

$$C = \frac{43.23^2}{(3)(3)(2)} = 103.82405.$$

$$SST = \sum_i\sum_j\sum_k y_{ijk}^2 - C = 104.43110 - 103.82405 = .60705$$

$$
\begin{aligned}
SS(Tr) &= \frac{1}{r}\sum_i\sum_j T_{ij.}^2 - C \\
&= \frac{1}{2}(4.09^2 + 4.64^2 + \cdots + 5.11^2 + 5.20^2) - 103.82405 = .55950 \\
SSR &= \frac{1}{ab}\sum_k T_{..k}^2 - C \\
&= \frac{1}{(3)(3)}(21.75^2 + 21.48^2) - 103.82405 = .00405 \\
SSE &= SST - SS(Tr) - SSR = .04350 \\
SSA &= \frac{1}{br}\sum_{i=1}^{a} T_{i..}^2 - C = \frac{1}{(3)(2)}(13.90^2 + 14.37^2 + 14.96^2) - 103.82405 \\
&= .09403 \\
SSB &= \frac{1}{ar}\sum_{j=1}^{b} T_{.j.}^2 - C = \frac{1}{(3)(2)}(13.37^2 + 14.36^2 + 15.50^2) - 103.82405 \\
&= .37870 \\
SS(AB) &= SS(Tr) - SSA - SSB = .08677
\end{aligned}
$$

The analysis-of-variance table is

Source of variation	Degrees of freedom	Sums of squares	Mean square	F
Replication	1	.00405	.00405	.74
A (Flow)	2	.09403	.04702	8.65
B (Filler)	2	.37870	.18935	34.82
AB interaction:	4	.08677	.02169	3.99
Error	8	.04350	.00544	
Total	17	.60705		

Since $F_{.05} = 5.32$ with 1 and 8 degrees of freedom, the replication effect is not significant at the .05 level. Since $F_{.01} = 8.65$ with 2 and 8 degrees of freedom, the Flow and Filler effects are both significant at .01 level. Since $F_{.05} = 3.84$ with 4 and 8 degrees of freedom, the Flow–Filter interaction effect is also significant at .05 level.

13.7 We have $a = 3$, $b = 2$, $c = 2$, $d = 3$, $r = 2$ and

$$T_{....1} = 1,059, \quad T_{....2} = 1,020, \quad T_{.....} = 2,079$$

$$\sum_{i=1}^{3}\sum_{j=1}^{2}\sum_{k=1}^{2}\sum_{m=1}^{3} T_{ijkm.}^2 = 127,787 \quad \text{and} \quad \sum_{i=1}^{3}\sum_{j=1}^{2}\sum_{k=1}^{2}\sum_{m=1}^{3}\sum_{l=1}^{2} y_{ijkml}^2 = 64,553$$

Hence $C = 2079^2/(3)(2)(2)(3)(2) = 60,031.125$.

$$
\begin{aligned}
SST &= 64553.000 - 60031.125 = 4521.875 \\
SS(Tr) &= \frac{1}{2}(127787.000) - 60031.125 = 3862.375
\end{aligned}
$$

$$\begin{aligned} SSR &= \frac{1}{(3)(2)(2)(3)}\left(1059^2 + 1020^2\right) - 60031.125 = 21.125 \\ SSE &= SST - SS(Tr) - SSR = 4521.875 - 3862.375 - 21.125 \\ &= 638.375 \end{aligned}$$

Next we construct all possible two-way tables:

		B: 1	B: 2	
A	Top	351	328	679
	Mid	348	362	710
	Bot	341	349	690
		1040	1039	2079

		C: Turn	C: Grind	
A	Top	357	322	679
	Mid	389	321	710
	Bot	358	332	690
		1104	975	2079

		D: 2100^0F	D: 2200^0F	D: 2300^0F	
A	Top	153	226	300	679
	Mid	181	246	283	710
	Bot	161	228	301	690
		495	700	884	2079

		C: Turn	C: Grind	
B	1	543	497	1040
	2	561	478	1039
		1104	975	2079

		D: 2100^0F	D: 2200^0F	D: 2300^0F	
B	1	244	352	444	1040
	2	251	348	440	1039
		495	700	884	2079

		D: 2100^0F	D: 2200^0F	D: 2300^0F	
C	Turn	272	360	472	1104
	Grind	223	340	412	975
		495	700	884	2079

From the tables, we have

$$\begin{aligned} SSA &= \frac{1}{bcdr}\sum_{i=1}^{a} T_{i....}^2 - C = \frac{1,441,241}{(2)(2)(3)(2)} - 60,031.125 \\ &= 20.583 \end{aligned}$$

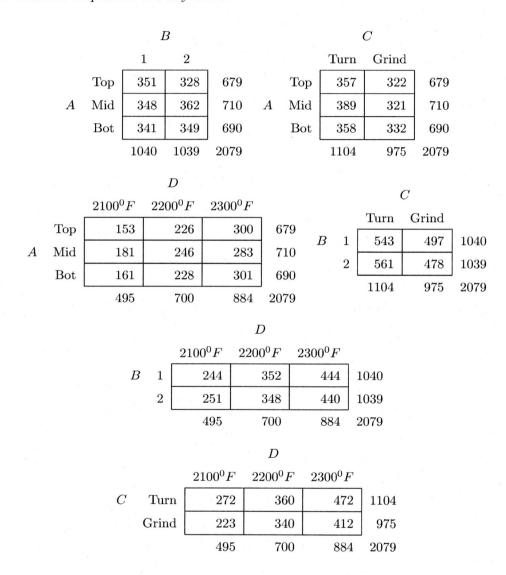

$$\begin{aligned}
SSB &= \frac{1}{acdr}\sum_{j=1}^{b} T^2_{.j...} - C = \frac{2,161,121}{(3)(2)(3)(2)} - 60,031.125 \\
&= .014 \\
SSC &= \frac{1}{abdr}\sum_{k=1}^{c} T^2_{..k..} - C = \frac{2,169,441}{(3)(2)(3)(2)} - 60,031.125 \\
&= 231.125 \\
SSD &= \frac{1}{abcr}\sum_{m=1}^{d} T^2_{...m.} - C = \frac{1,516,481}{(3)(2)(2)(2)} - 60,031.125 \\
&= 3,155.583
\end{aligned}$$

Since

$$\frac{1}{cdr}\sum_{i=1}^{a}\sum_{j=1}^{b} T^2_{ij...} - C = \frac{721,015}{(2)(3)(2)} - 60,031.125 = 53.458$$

$$\frac{1}{bdr}\sum_{i=1}^{a}\sum_{k=1}^{c} T^2_{i.k..} - C = \frac{732,883}{(2)(3)(2)} - 60,031.125 = 292.458$$

$$\frac{1}{bcr}\sum_{i=1}^{a}\sum_{m=1}^{d} T^2_{i..m.} - C = \frac{506,357}{(2)(2)(2)} - 60,031.125 = 3,263.500$$

$$\frac{1}{adr}\sum_{j=1}^{b}\sum_{k=1}^{c} T^2_{.jk..} - C = \frac{1,085,063}{(3)(3)(2)} - 60,031.125 = 250.153$$

$$\frac{1}{acr}\sum_{j=1}^{b}\sum_{m=1}^{d} T^2_{.j.m.} - C = \frac{758,281}{(3)(2)(2)} - 60,031.125 = 3,158.960$$

$$\frac{1}{abr}\sum_{k=1}^{c}\sum_{m=1}^{d} T^2_{..km.} - C = \frac{761,441}{(3)(2)(2)} - 60,031.125 = 3,422.292$$

we have

$$\begin{aligned}
SS(AB) &= 53.458 - 20.583 - .014 = 32.861 \\
SS(AC) &= 292.458 - 20.583 - 231.125 = 40.750 \\
SS(AD) &= 3,263.500 - 20.583 - 3,155.583 = 87.334 \\
SS(BC) &= 250.153 - .014 - 231.125 = 19.014 \\
SS(BD) &= 3,158.960 - .014 - 3,155.583 = 3.361 \\
SS(CD) &= 3,422.292 - 231.125 - 3,155.583 = 35.583
\end{aligned}$$

Since

$$\frac{1}{dr}\sum_{i=1}^{a}\sum_{j=1}^{b}\sum_{k=1}^{c} T^2_{ijk..} - C = \frac{362,313}{(3)(2)} - 60,031.125 = 354.375$$

$$\frac{1}{cr}\sum_{i=1}^{a}\sum_{j=1}^{b}\sum_{m=1}^{d}T_{ij.m.}^2 - C = \frac{253,477}{(2)(2)} - 60,031.125 = 3,338.125$$

$$\frac{1}{br}\sum_{i=1}^{a}\sum_{k=1}^{c}\sum_{k=1}^{d}T_{i.km.}^2 - C = \frac{254,921}{(2)(2)} - 60,031.125 = 3,699.125$$

$$\frac{1}{ar}\sum_{j=1}^{b}\sum_{k=1}^{c}\sum_{m=1}^{d}T_{.jkm.}^2 - C = \frac{380,913}{(3)(2)} - 60,031.125 = 3,454.375$$

we have

$$\begin{aligned}
SS(ABC) &= 354.375 - SSA - SSB - SSC - SS(AB) - SS(AC) \\
&\quad -SS(BC) = 10.028 \\
SS(ABD) &= 3,338.125 - SSA - SSB - SSD - SS(AB) - SS(AD) \\
&\quad -SS(BD) = 38.389 \\
SS(ACD) &= 3,699.125 - SSA - SSC - SSD - SS(AC) - SS(AD) \\
&\quad -SS(CD) = 128.167 \\
SS(BCD) &= 3,454.375 - SSB - SSC - SSD - SS(BC) - SS(BD) \\
&\quad -SS(CD) = 9.693 \\
SS(ABCD) &= SS(Tr) - SSA - SSB - SSC - SSD - SS(AB) \\
&\quad -SS(AC) - SS(AD) - SS(BC) - SS(BD) - SS(CD) \\
&\quad -SS(ABC) - SS(ABD) - SS(ACD) - SS(BCD) \\
&= 49.890
\end{aligned}$$

The analysis-of-variance table is

Source of variation	Degrees of freedom	Sums of squares	Mean square	F
Replication	1	21.125	21.125	1.158
Main effects:				
A	2	20.583	10.291	.564
B	1	.014	.014	.001
C	1	231.125	231.125	12.672
D	2	3,155.583	1,577.792	86.506
Two-factor interactions:				
AB	2	32.861	16.431	.901
AC	2	40.750	20.375	1.117
AD	4	87.334	21.834	1.197
BC	1	19.014	19.014	1.042
BD	2	3.361	1.681	.092
CD	2	35.583	17.792	.975

table continued

Source of variation	Degrees of freedom	Sums of squares	Mean square	F
Three-factor interactions:				
ABC	2	10.028	5.014	.275
ABD	4	38.389	9.597	.526
ACD	4	128.167	32.042	1.757
BCD	2	9.693	4.847	.266
$ABCD$ interaction:	4	49.890	12.473	.684
Error	35	638.375	18.239	
Total	71	4,521.875		

At the .05 level, the critical values are

Degrees of freedom	F
1 and 35	4.12
2 and 35	3.27
4 and 35	2.64

Thus main effects C(specimen preparation) and D(twisting temperature) are significant at the .05 level. No other effects are significant at the .05 level.

From Figure 13.1, it is clear that the smallest number of turns to break the steel occurred at $2,100^0 F$ for steel that had undergone grinding.

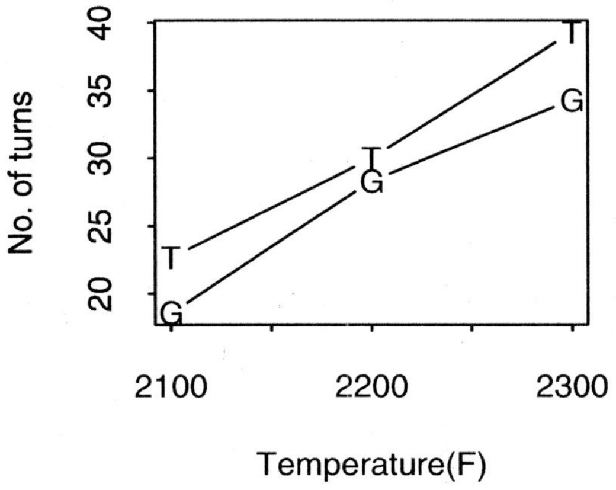

Figure 13.1: T and G stand for Turn and Grind, respectively. Exercise 13.7.

13.9 The four equations are:

$$\mu_{111} = \mu + \alpha_1 + \beta_1 + (\alpha\beta)_{11}$$
$$\mu_{121} = \mu + \alpha_1 - \beta_1 - (\alpha\beta)_{11}$$
$$\mu_{211} = \mu - \alpha_1 + \beta_1 - (\alpha\beta)_{11}$$
$$\mu_{221} = \mu - \alpha_1 - \beta_1 + (\alpha\beta)_{11}$$

Now solve the equations by combining the four population means on the left hand sides. We first sum the means.

$$\tfrac{1}{4}(\mu_{111} + \mu_{121} + \mu_{211} + \mu_{221}) = \mu$$
$$\tfrac{1}{4}(\mu_{111} + \mu_{121} - \mu_{211} - \mu_{221}) = \alpha_1$$
$$\tfrac{1}{4}(\mu_{111} - \mu_{121} + \mu_{211} - \mu_{221}) = \beta_1$$
$$\tfrac{1}{4}(\mu_{111} - \mu_{121} - \mu_{211} + \mu_{221}) = (\alpha\beta)_{11}$$

13.11 (a) The analysis-of-variance table for the two-way classification with 7 degrees of freedom for treatments and 1 degree of freedom for replicates is

Source of variation	Degrees of freedom	Sums of squares	Mean square	F
Replication	1	1.5625	1.5625	2.22
Treatments	7	103.9375	14.8482	21.05
Error	7	4.9375	.7054	
Total	15	110.4375		

Since $F_{.05} = 5.59$ with 1 and 7 degrees of freedom, the replication effect is not significant at the .05 level. Since $F_{.01} = 6.99$ with 7 and 7 degrees of freedom, the treatment effect is significant at the .01 level.

(b) We can use the table of signs on page 411. The data written in standard order are

	A Lubricant	*B* Heat	*C* Resin	Ratings Rep. 1	Ratings Rep. 2	Total
1	Fresh	Unheated	*A*	6	8	14
a	Aged	Unheated	*A*	6	7	13
b	Fresh	Heated	*A*	9	9	18
ab	Aged	Heated	*A*	9	8	17
c	Fresh	Unheated	*B*	8	7	15
ac	Aged	Unheated	*B*	6	8	14
bc	Fresh	Heated	*B*	1	2	3
abc	Aged	Heated	*B*	2	3	5

Thus, from the table of signs, the effect totals are

$$
\begin{aligned}
[I] &= 14+13+18+17+15+14+3+5 \;=\; 99\\
[A] &= -14+13-18+17-15+14-3+5 \;=\; -1\\
[B] &= -14-13+18+17-15-14+3+5 \;=\; -13\\
[AB] &= 14-13-18+17+15-14-3+5 \;=\; 3\\
[C] &= -14-13-18-17+15+14+3+5 \;=\; -25\\
[AC] &= 14-13+18-17-15+14-3+5 \;=\; 3\\
[BC] &= 14+13-18-17-15-14+3+5 \;=\; -29\\
[ABC] &= -14+13+18-17+15-14-3+5 \;=\; 3
\end{aligned}
$$

(c) Since $n = 3$ and $r = 2$, we have

$SSA = (-1)^2/2(2^2) = .0625$ $\quad$ $SSB = (-13)^2/2(2^2) = 10.0625$

$SSC = (-25)^2/2(2^2) = 39.0625$ $\quad$ $SSAB = (3)^2/2(2^2) = .5625$

$SSAC = (3)^2/2(2^2) = .5625$ $\quad$ $SSBC = (-29)^2/2(2^2) = 52.5625$

$SSABC = (3)^2/2(2^2) = .5625$

To check that these sums of squares add up to $SS(Tr)$ from part (a), we have

$$.0625 + 10.5625 + 39.0625 + .5625 + .5625 + 52.5625 + .5625 = 103.9375$$

as required.

(d) The data were arranged in standard order in part (b). The table of calculations using the Yates method is

Exp. con.	Tr. total	1	2	3	Sum of squares
1	14	27	62	99	612.5625
a	13	35	37	-1	.0625
b	18	29	-2	-13	10.5625
ab	17	8	1	3	.5625
c	15	-1	8	25	39.0625
ac	14	-1	-21	3	.5625
bc	3	-1	0	-29	52.5625
abc	5	2	3	3	.5625

These numbers match those of parts (b) and (c).

(e) The analysis-of-variance table is

Source of variation	Degrees of freedom	Sums of squares	Mean square	F
Replication	1	1.5625	1.5625	2.2151
Main effects:				
A	1	.0625	.0625	.0886
B	1	10.5625	10.5625	14.9738
C	1	39.0625	39.0625	55.3764
Two-factor interactions:				
AB	1	.5625	.5625	.7974
AC	1	.5625	.5625	.7974
BC	1	52.5625	52.5625	74.5145
ABC interaction:	1	.5625	.5625	.7974
Error	7	4.9375	.7054	
Total	15	110.4375		

The critical value is $F_{.05} = 5.59$ with 1 and 7 degrees of freedom. The effects for B, C and BC interaction are significant at the .05 level. We don't interpret main effects individually because of the apparent BC interaction.

13.13 The table of calculations using the Yates method is given in the first table. The table for analysis-of-variance follows.

Yates method for Exercise 13.13						
Exp. con.	Tr. total	1	2	3	4	Sum of squares
1	82.2	157.7	337.3	690.6	1376.6	59,219.6113
a	75.5	179.6	353.3	686.0	−44.2	61.0513
b	98.4	166.5	352.5	−5.2	92.6	267.9612
ab	81.2	186.8	333.5	−39.0	25.8	20.8013
c	84.3	164.5	−23.9	42.2	−3.0	.2813
ac	82.2	188.0	18.7	50.4	58.2	105.8513
bc	83.0	153.3	−27.3	12.4	1.8	.1013
abc	103.8	180.2	−11.7	13.4	76.6	183.3612
d	82.0	−6.7	21.9	16.0	−4.6	.6613
ad	82.5	−17.2	20.3	−19.0	−33.8	35.7013
bd	107.9	−2.1	23.5	42.6	8.2	2.1013
abd	80.1	20.8	26.9	15.6	1.0	.0313
cd	83.3	.5	−10.5	−1.6	−35.0	38.2813
acd	70.0	−27.8	22.9	3.4	−27.0	22.7813
bcd	89.3	−13.3	−28.3	33.4	5.0	.7813
$abcd$	90.9	1.6	14.9	43.2	9.8	3.0013

The analysis-of-variance table is

Source of variation	Degrees of freedom	Sums of squares	Mean square	F
Replication	1	300.1250	300.1250	9.0787
Main effects:				
A	1	61.0513	61.0513	1.8468
B	1	267.9612	267.9612	8.1057
C	1	.2813	.2813	.0085
D	1	.6613	.6613	.0200
Two-factor interactions:				
AB	1	20.8013	20.8013	.6292
AC	1	105.8513	105.8513	3.2020
AD	1	35.7013	35.7013	1.0799
BC	1	.1013	.1013	.0030
BD	1	2.1013	2.1013	.0636
CD	1	38.2813	38.2813	1.1580
Three-factor interactions:				
ABC	1	183.3612	183.3612	5.5466
ABD	1	.0313	.0313	.0091
ACD	1	22.7813	22.7813	.6891
BCD	1	.7813	.7813	.0236
$ABCD$ interaction:	1	3.0013	3.0013	.0908
Error	15	495.8750	33.0583	
Total	31	1,538.7487		

The critical value is $F_{.05} = 4.54$ with 1 and 15 degrees of freedom. The replication effect is significant at the .05 level and so are the main effect of B and the ABC interaction. It seems that aging time has no effect on the gain of the semiconductor device.

13.15 From the model

$$y_{100l} = \mu + \alpha_1 + \beta_0 + \gamma_0 + (\alpha\beta)_{10} + (\alpha\gamma)_{10} + (\beta\gamma)_{00} + (\alpha\beta\gamma)_{100} + \rho_l + \epsilon_{100l}$$

Thus,

$$\sum_{l=1}^{r} y_{100l} = r\,(\,\mu + \alpha_1 + \beta_0 + \gamma_0 + (\alpha\beta)_{10} + (\alpha\gamma)_{10}$$

$$+(\beta\gamma)_{00} + (\alpha\beta\gamma)_{100}\,) + \sum_{l=1}^{r} \rho_l + \sum_{l=1}^{r} \epsilon_{100l}$$

Since $\sum_{l=1}^{r} \rho_l = 0$ and $(a) = \sum_{l=1}^{r} y_{100l}$, we have

$$(a) = r[\mu + \alpha_1 + \beta_0 + \gamma_0 + (\alpha\beta)_{10} + (\alpha\gamma)_{10}$$

$$+(\beta\gamma)_{00} + (\alpha\beta\gamma)_{100}] + \sum_{l=1}^{r} \epsilon_{100l}$$

But, the constraints are $\alpha_1 = -\alpha_0$,

$$(\alpha\beta)_{10} = -(\alpha\beta)_{00}\ , \quad (\alpha\gamma)_{10} = -(\alpha\gamma)_{00}\ , \quad (\alpha\beta\gamma)_{100} = -(\alpha\beta\gamma)_{000}$$

so that

$$(a) = r[\mu - \alpha_0 + \beta_0 + \gamma_0 - (\alpha\beta)_{00} - (\alpha\gamma)_{00} + (\beta\gamma)_{00} - (\alpha\beta\gamma)_{000}] + \sum_{l=1}^{r} \epsilon_{100l}$$

13.17 Multiplying each of the equations in the previous exercise by the appropriate sign and adding gives

$$[A] = -(1) + (a) - (b) + (ab) - (c) + (ac) - (bc) + (abc) = -8r\alpha_0 + \epsilon_A$$

where

$$\epsilon_A = \sum_{l=1}^{r}(-\epsilon_{000l} + \epsilon_{100l} - \epsilon_{010l} + \epsilon_{110l} - \epsilon_{001l} + \epsilon_{101l} - \epsilon_{101l} + \epsilon_{111l})$$

and

$$[BC] = (1) + (a) - (b) - (ab) - (c) - (ac) + (bc) + (abc) = 8r(\beta\gamma)_{00} + \epsilon_{BC}$$

where

$$\epsilon_{BC} = \sum_{l=1}^{r}(\epsilon_{000l} + \epsilon_{100l} - \epsilon_{010l} - \epsilon_{110l} - \epsilon_{001l} - \epsilon_{101l} + \epsilon_{101l} + \epsilon_{111l})$$

13.19 Summing up the various sum of squares, we have

$$134,551 + 4,632 + 225,624 + 713 + 39,410 + 18,673 + 31,689 + 1,001$$

$$+24,698 + 81 + 39,130 + 30 + 12,601 + 14,070 + 385 = 547,228$$

This is the same as $SS(Tr)$ obtained in the calculations leading to the table.

13.21 (a) We verify the calculation of mean square using the formulas based on totals.

$$T_{1.} = 24, T_{2.} = 40, T_{3.} = 32, T_{4.} = 44, \text{and } \ T_{..} = 40$$

so $C = 140^2/8 = 2,450$. Further

$$T_{.1} = 68, T_{.2} = 72, \text{and } \ \sum\sum y_{ij}^2 = 2,588$$

so

$$\begin{aligned} SST &= 2,588 - 2,450 = 138 \\ SS(Tr) &= \frac{1}{2}(5,136) - 2,450 = 118 \end{aligned}$$

Therefore $SSE = 138 - 118 = 20$ with 4 degrees of freedom and $MSE = 20/4 = 5$ which verifies the result in the example.

(b)

$$\frac{[B]}{2r} = \frac{1}{2r}[-(1) - (a) + (b) + (ab)]$$

$$= \frac{1}{2r}[-\sum_{k=1}^{r} y_{00k} - \sum_{k=1}^{r} y_{10k} + \sum_{k=1}^{r} y_{01k} + \sum_{k=1}^{r} y_{11k}]$$

$$= \frac{1}{2r}[-\sum_{i=0}^{1}\sum_{k=1}^{r} y_{i0k} + \sum_{i=0}^{1}\sum_{k=1}^{r} y_{i1k}] = \bar{y}_{.1.} - \bar{y}_{.0.}$$

(c)

$$\frac{[AB]}{2r} = \frac{1}{2r}[(1) - (a) - (b) + (ab)]$$

$$= \frac{1}{2r}[\sum_{k=1}^{r} y_{00k} - \sum_{k=1}^{r} y_{10k} - \sum_{k=1}^{r} y_{01k} + \sum_{k=1}^{r} y_{11k}]$$

$$= \frac{1}{2}[\frac{1}{r}\sum_{k=1}^{r} y_{00k} - \frac{1}{r}\sum_{k=1}^{r} y_{10k} - \frac{1}{r}\sum_{k=1}^{r} y_{01k} + \frac{1}{r}\sum_{k=1}^{r} y_{11k}]$$

$$= \frac{1}{2}(\bar{y}_{11.} + \bar{y}_{00.}) - \frac{1}{2}(\bar{y}_{10.} + \bar{y}_{01.})$$

13.23 The visual summary of the four treatment means is given in Figure 13.2. To obtain confidence intervals

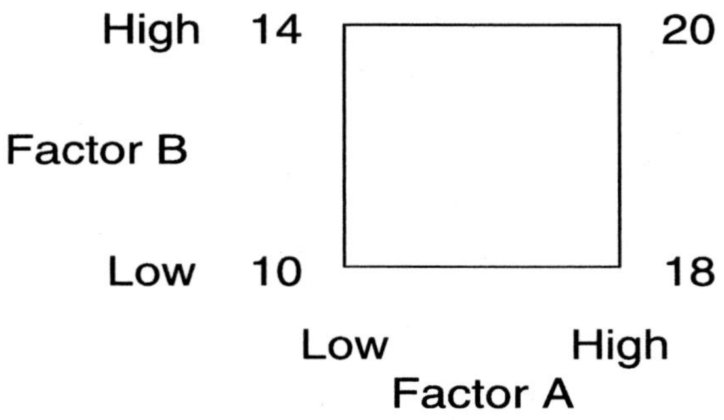

Figure 13.2: Visual summary of the experiment. Exercise 13.23

for the mean effects, we calculate

$$\widehat{\alpha_1 - \alpha_0} = \bar{y}_{1..} - \bar{y}_{0..} = 19 - 12 = 7$$

$$\widehat{\beta_1 - \beta_0} = \bar{y}_{.1.} - \bar{y}_{.0.} = 17 - 14 = 3$$

$$\widehat{\text{interaction}} = \frac{1}{2}(\bar{y}_{11.} - \bar{y}_{10.} - \bar{y}_{01.} + \bar{y}_{00.})$$

$$= \frac{1}{2}(20 - 18 - 14 + 10) = -1$$

Next, we find the mean square error.

$$\begin{aligned} C &= 124^2/8 = 1,922 \\ SST &= 2,140 - 1,922 = 218 \\ SS(Tr) &= \tfrac{1}{2}(4,080) - 1,922 = 118 \end{aligned}$$

so $SSE = 100$ with 4 degrees of freedom and $s^2 = MSE = 100/4 = 25$. Because $t_{.025} = 2.776$ for 4 degrees of freedom, the confidence intervals are

Factor A :

$$\bar{y}_{1..} - \bar{y}_{0..} \pm t_{.025}\sqrt{\frac{s^2}{r}} = 7 \pm (2.776)\sqrt{\frac{25}{2}} = 7 \pm 9.8 \text{ or } -2.8 \text{ to } 16.8$$

Factor B :

$$\bar{y}_{.1.} - \bar{y}_{.0.} \pm t_{.025}\sqrt{\frac{s^2}{r}} = 7 \pm (2.776)\sqrt{\frac{25}{2}} = 3 \pm 9.8 \text{ or } -6.8 \text{ to } 12.8$$

Interaction AB :

$$-1 \pm 9.8 \text{ or } -10.8 \text{ to } 8.8$$

All of the intervals cover 0 so none of the effects are significant. The experiment can, ultimately, be summarized by the common mean $\bar{y}_{...} = 124/8 = 15.5$ and $s = 5$.

13.25 (a) The data and replicate means are

	Rep. 1	Rep. 2	Mean	$y_{ijk1} - \bar{y}_{ijk.}$	$y_{ijk2} - \bar{y}_{ijk.}$
1	4.5	4.1	4.3	.2	−.2
a	3.8	3.4	3.6	.2	−.2
b	3.1	4.3	3.7	−.6	.6
ab	7.2	6.8	7.0	.2	−.2
c	5.4	5.0	5.2	.2	−.2
ac	4.5	4.9	4.7	−.2	.2
bc	4.2	5.4	4.8	−.6	.6
abc	7.3	6.9	7.1	.2	−.2

Note that $y_{ijk1} - \bar{y}_{ijk.} = -(y_{ijk2} - \bar{y}_{ijk.})$. Squaring the entries in the last two columns and summing, or taking twice the sum of squares for one column, we obtain

$$SSE = 2(.2^2 + .2^2 + \cdots + .2^2) - 2^3[(-.05)^2 + (.05)^2] = 1.92 - .04 = 1.88$$

Therefore, $s^2 = 1.88/7 = .2686$

(b) Since $t_{.025} = 2.365$ for 7 degrees of freedom, the half length of the confidence intervals is

$$t_{.025}\sqrt{\frac{s^2}{2r}} = 2.365\sqrt{\frac{.2686}{4}} = 0.61.$$

Replacing the half length of the confidence intervals shown in the example by .61, the resulting individual 95% confidence intervals are:

rate effect: $1.1 \pm .61$ or $.49$ to 1.71,

additive effect: $1.2 \pm .61$ or $.59$ to 1.81,

nozzle effect: $.8 \pm .61$ or $.19$ to 1.41,

rate $\times$ additive interaction: $1.7 \pm .61$ or 1.09 to 2.31,

rate $\times$ nozzle interaction: $-.2 \pm .61$ or $-.81$ to $.41$,

additive $\times$ nozzle interaction: $-.2 \pm .61$ or $-.81$ to $.41$,

rate $\times$ additive $\times$ nozzle interaction: $-.3 \pm .61$ or $-.91$ to $.31$.

13.27 The visual summary of the eight treatment means is given in Figure 13.3. According to visual procedure, we assume that there is no replication effects in the model. The error sum of squares is

$$SSE = \sum_{i=0}^{1}\sum_{j=0}^{1}\sum_{k=0}^{1}\sum_{l=1}^{r}(y_{ijkl} - \bar{y}_{ijk.})^2$$

with $2^3(r-1)$ degrees of freedom. Thus we calculate

	Rep. 1	Rep. 2	Mean	$y_{ijk1} - \bar{y}_{ijk.}$	$y_{ijk2} - \bar{y}_{ijk.}$
1	3.7	4.1	3.9	−.2	.2
a	4.6	5.0	4.8	−.2	.2
b	3.1	2.7	2.9	.2	−.2
ab	3.4	3.8	3.6	−.2	.2
c	3.4	3.6	3.5	−.1	.1
ac	5.3	4.9	5.1	.2	−.2
bc	2.4	3.2	2.8	−.4	−.4
abc	4.7	4.1	4.4	.3	−.3
Mean	3.825	3.925	3.875		

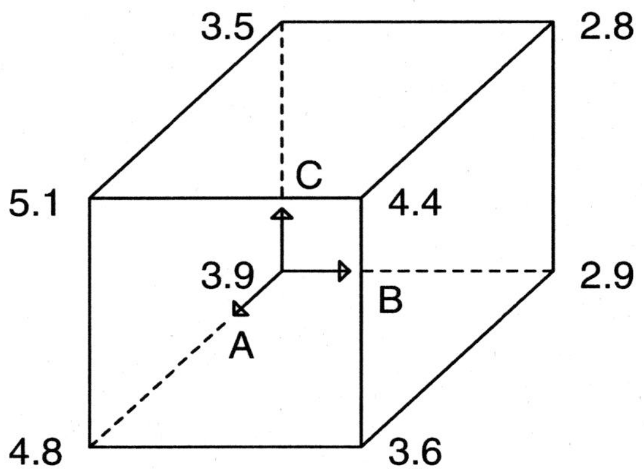

Figure 13.3: Visual summary of the experiment. Exercise 13.27

Note that $y_{ijk1} - \bar{y}_{ijk.} = -(y_{ijk2} - \bar{y}_{ijk.})$. Squaring the entries in the last two columns and summing, or taking twice the sum of squares for one column, we obtain

$$SSE = 2((-.2)^2 + (-.2)^2 + \cdots + .3^2) = .92$$

with $2^3(2-1) = 8$ degrees of freedom. Therefore, $s^2 = MSE = 92/8 = .115$. Since $t_{.025} = 2.306$ for 8 degrees of freedom, the half length of the confidence intervals is

$$t_{.025}\sqrt{\frac{s^2}{2r}} = 2.306\sqrt{\frac{.115}{4}} = .39$$

The estimated effects are:

Factor A :

$$\bar{y}_{1...} - \bar{y}_{0...} = 4.475 - 3.275 = 1.20$$

Factor B :

$$\bar{y}_{.1..} - \bar{y}_{.0..} = 3.425 - 4.325 = -.90$$

Factor C :

$$\bar{y}_{..1.} - \bar{y}_{..0.} = 3.950 - 3.800 = .150$$

AB :

$$\frac{1}{2}(\bar{y}_{11..} - \bar{y}_{10..} - \bar{y}_{01..} + \bar{y}_{00..}) = \frac{1}{2}(4.0 - 4.95 - 2.85 + 3.7) = -.05$$

AC :

$$\frac{1}{2}(\bar{y}_{1.1.} - \bar{y}_{1.0.} - \bar{y}_{0.1.} + \bar{y}_{0.0.}) = \frac{1}{2}(4.75 - 4.2 - 3.15 + 3.4) = .4$$

BC :

$$\frac{1}{2}(\bar{y}_{.11.} - \bar{y}_{.10.} - \bar{y}_{.01.} + \bar{y}_{.00.}) = \frac{1}{2}(3.6 - 3.25 - 4.3 + 4.35) = .2$$

ABC :

$$\frac{1}{4}(\bar{y}_{111.} - \bar{y}_{101.} - \bar{y}_{011.} + \bar{y}_{001.}) - \frac{1}{4}(\bar{y}_{110.} - \bar{y}_{100.} - \bar{y}_{010.} + \bar{y}_{000.})$$

$$= \frac{1}{4}(4.4 - 5.1 - 2.8 + 3.5) - \frac{1}{4}(3.6 - 4.8 - 2.9 + 3.9) = .05$$

so the confidence intervals are:

(Viscosity) A :

$$1.2 \pm .39 \quad \text{or} \quad .81 \text{ to } 1.59$$

(Temperature) B :

$$-.9 \pm .39 \quad \text{or} \quad -1.29 \text{ to } -.51$$

(Additive) C :

$$.15 \pm .39 \quad \text{or} \quad -.24 \text{ to } .54$$

AB :

$$-.05 \pm .39 \quad \text{or} \quad -.44 \text{ to } .34$$

AC :

$$.4 \pm .39 \quad \text{or} \quad .01 \text{ to } .79$$

BC :

$$.2 \pm .39 \quad \text{or} \quad -.19 \text{ to } .59$$

ABC :

$$.05 \pm .39 \quad \text{or} \quad -.34 \text{ to } .44$$

Only the confidence intervals for the main effects of oil viscosity and temperature and the viscosity $\times$ additive interaction fail to cover 0. Over the conditions of this experiment, low temperature produces a higher mean for the response.

13.29 Similar to Exercise 13.21, it can be shown that in a 2^n design, for each treatment,

$$\text{treatment total} = (\text{estimated effect}) \times r2^{n-1}$$

Thus the sum of squares of a treatment is

$$\begin{aligned} \text{sum of squares} &= \frac{(\text{treatment total})^2}{r2^n} \\ &= \frac{\left[(\text{estimated effect}) \times r2^{n-1}\right]^2}{r2^n} \\ &= (\text{estimated effect})^2 \times r2^{n-2} \end{aligned}$$

Since $n = 3$ and $r = 2$, using the estimated effects given in the example, we calculate

Treatment	Estimated effect	Sum of squares
A	1.1	$1.1^2(2)(2) = 4.84$
B	1.2	$1.2^2(2)(2) = 5.76$
C	.8	$.8^2(2)(2) = 2.56$
AB	1.7	$1.7^2(2)(2) = 11.56$
AC	$-.2$	$(-.2)^2(2)(2) = .16$
BC	$-.2$	$(-.2)^2(2)(2) = .16$
ABC	$-.3$	$(-.3)^2(2)(2) = .36$

From the example, we also know that $MSE = s^2 = .24$ with $2^3(2-1) = 8$ degrees of freedom. Hence the analysis-of-variance table is

Source of variation	Degrees of freedom	Sums of squares	Mean square	F
Main effects:				
A Rate	1	4.84	4.84	20.17
B Additive	1	5.76	5.76	24.00
C Nozzle	1	2.56	2.56	10.67
Two-factor interactions:				
AB	1	11.56	11.56	48.17
AC	1	.16	.16	.67
BC	1	.16	.16	.67
ABC interaction:	1	.36	.36	1.50
Error	8	1.92	.24	
Total	15	27.32		

Since $F_{.01} = 11.26$ with 1 and 8 degrees of freedom, the rate and additive effects and their interaction are significant at the .01 level. Since $F_{.05} = 5.32$ with 1 and 8 degrees of freedom, the nozzle effect is significant at the .05 level. No other effects are significant at the .05 level.

13.31 The response surface analysis changes only slightly. especially the squares and product term.

```
Response Surface Regression: Yield versus Additive, Temperature


The analysis was done using uncoded units.

Estimated Regression Coefficients for Yield

Term                          Coef  SE Coef        T      P
Constant                   52.5417  17.9237    2.931  0.043
Additive                    0.0798   0.1435    0.556  0.608
Temperature                 0.6521   0.2614    2.495  0.067
Additive*Additive          -0.0147   0.0012  -12.170  0.000
Temperature*Temperature    -0.0038   0.0009   -4.057  0.015
Additive*Temperature        0.0073   0.0008    9.074  0.001

S = 2.25924
R-Sq = 98.74%

Analysis of Variance for Yield
Source          DF   Seq SS    Adj SS    Adj MS      F      P
Regression       5  1605.98  1605.983  321.197  62.93  0.001
  Linear         2   235.33    33.520   16.760   3.28  0.143
  Square         2   950.40   950.400  475.200  93.10  0.000
  Interaction    1   420.25   420.250  420.250  82.33  0.001
Residual Error   4    20.42    20.417    5.104
  Lack-of-Fit    3    19.92    19.917    6.639  13.28  0.198
  Pure Error     1     0.50     0.500    0.500
Total            9  1626.40
```

13.33 The contour plot in Figure 13.4 shows that there is a region in the lower right hand corner where the predicted adhesion is greater than 45 grams. It is different from the region where yield is maximized.

13.35 (a) Since,

$$(a-1)(b-1)(c-1)(d-1)$$

$$= abcd - acd - bcd + cd - abd + ad + bd - d$$

$$-abc + ac + bc - c + ab - a - b + 1$$

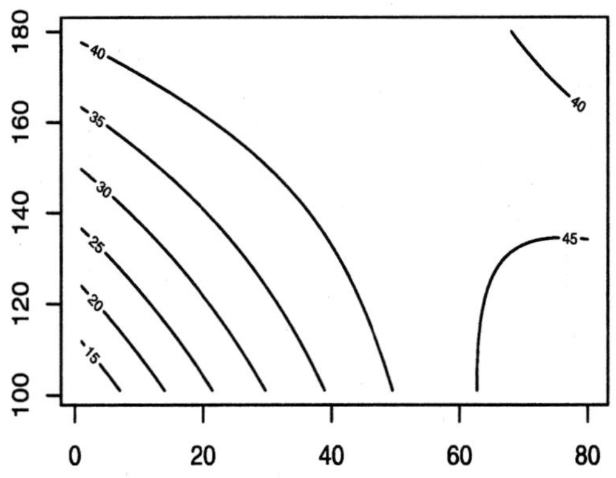

Figure 13.4: Contour plot of adhesion. Exercise 13.33

we can use the blocks

$$Block\ 1 : 1, ab, ac, bc, bd, ad, cd, abcd$$

$$Block\ 2 : a, b, c, abc, d, abd, bcd, acd$$

(b) Since

$$(a-1)(b-1)(c-1)(d+1)$$

$$= abcd - acd - bcd + cd - abd + ad + bd - d$$

$$+abc - ac - bc + c - ab + a + b - 1$$

we first divide into two blocks

$$Block\ 1 : 1, ab, ac, bc, d, abd, bcd, acd$$

$$Block\ 2 : a, b, c, abc, bd, ad, cd, abcd$$

Since

$$(a+1)(b-1)(c-1)(d-1)$$

$$= abcd - acd + bcd - cd - abd + ad - bd + d$$

$$-abc + ac - bc + c + ab - a + b - 1$$

we divide the two blocks above to get the required 4 blocks.

$$Block\ 1: 1, bc, abd, acd$$

$$Block\ 2: ab, ac, d, bcd$$

$$Block\ 3: a, bd, cd, abc$$

$$Block\ 4: b, ad, c, abcd$$

13.37 We first combine the data as if there are no confounding effects presented with the two replicates and calculate the treatment sum of squares using the Yates method.

Yates method for Exercise 13.37

Exp. con.	Tr. total	1	2	3	4	Sum of squares
1	4.6	10.0	22.8	52.6	101.0	312.7813
a	5.4	12.8	29.8	48.4	3.0	.2813
b	5.6	13.2	19.0	6.2	11.4	4.0613
ab	7.2	16.6	29.4	−3.2	−5.8	1.0513
c	4.2	8.2	2.4	6.2	17.4	9.4613
ac	9.0	10.8	3.8	5.2	5.0	.7813
bc	8.8	13.4	−3.4	−5.0	.6	.0113
abc	7.8	16.0	.2	−.8	.2	.0013
d	4.0	.8	2.8	7.0	−4.2	.5513
ad	4.2	1.6	3.4	10.4	−9.4	2.7613
bd	7.2	4.8	2.6	1.4	−1.0	.0313
abd	3.6	−1.0	2.6	3.6	4.2	.5513
cd	7.4	.2	.8	.6	3.4	.3613
acd	6.0	−3.6	−5.8	0.0	2.2	.1513
bcd	7.2	−1.4	−3.8	−6.6	−.6	.0113
$abcd$	8.8	1.6	3.0	6.8	13.4	5.6113

The table of block sums is

	Block 1	Block 2
Trial 1	28.8	28.4
Trial 2	23.2	20.6

Since $C = 318.7813$, we have

$$SST = 349.6400 - 318.7813 = 30.8588$$
$$SS(Tr) = 688.9200/2 - 318.7813 = 25.6788$$
$$SS(Bl) = 2598.6000/8 - 318.7813 = 6.0438$$

The analysis-of-variance table is

Source of variation	Degrees of freedom	Sums of squares	Mean square	F
Blocks	3	6.04375	2.0146	5.94
Main effects:				
A	1	.2813	.2813	.83
B	1	4.0613	4.0613	11.98
C	1	9.4613	9.4613	27.90
D	1	.5513	.5513	1.63
Two-factor interactions:				
AB	1	1.0513	1.0513	3.10
AC	1	.7813	.7813	2.30
AD	1	2.7613	2.7613	8.14
BC	1	.0113	.0113	.03
BD	1	.0313	.0313	.09
CD	1	.3613	.3613	1.07
Three-factor interactions:				
ABC	1	.0013	.0013	.00
ABD	1	.5513	.5513	1.63
ACD	1	.1513	.1513	.45
BCD	1	.0113	.0113	.03
Intrablock error	14	4.7475	.3391	
Total	31	30.8588		

The critical value is $F_{.05} = 4.60$ with 1 and 14 degrees of freedom. The effects for blocks, factors B and C and AD interaction are significant at the .05 level.

We summarize the findings in the following:

Tranq. B		Tranq. C		Tranq.
High	Low	High	Low	B High and C High
3.5	2.8	3.7	2.6	4.08

		Tranq. D	
		High	Low
Tranq. A	High	2.8	3.7
	Low	3.2	2.9

Running tranquilizers B and C at high doses level gives the best result since the effects of B and C are additive. It may also be advantagous to run A high simultaneously with D low.

13.39 The method of Exercise 13.20 allows the sign of any treatment total to be determined in the sum for any effect total. The odd-even method allows separation of the treatment totals according to sign. Thus, we most show that all "evens" have the same sign and all "odds" have the opposite sign. But the sign for a treatment total in the expansion of the appropriate product is -1 raised to the number of letters in the effect that are not in the treatment total. For example, for effect AB , the sign of (acd) is -1 since there is one letter (B) in AB that is not in (acd). Thus, "evens" must all have the same signs, and "odds" must have the opposite sign.

13.41 The experimental conditions are

$$1, ab, ac, ad, ae, af, bc, bd, be, bf, cd, ce, cf, de, df, ef,$$

$$abcd, abce, abcf, abde, abdf, abef, acde, acdf, acef,$$

$$adef, bcde, bcdf, bcef, bdef, cdef, abcdef,$$

The alias pairs are

A and BCDEF	E and ABCDF
B and ACDEF	AE and BCDF
AB and CDEF	BE and ACDF
C and ABDEF	ABE and CDF
AC and BDEF	CE and ABDF
BC and ADEF	ACE and BDF
ABC and DEF	BCE and ADF
D and ABCEF	ABCE and DF
AD and BCEF	DE and ABCF
BD and ACEF	ADE and BCF
ABD and CEF	BDE and ACF
CD and ABEF	ABDE and CF
ACD and BEF	CDE and ABF
BCD and AEF	ACDE and BF
ABCD and EF	BCEE and AF
	ABCDE and F

13.43 (a) The modified standard order is:

$$1, af, bf, ab, cf, ac, bc, abcf, df, ad, bd, abdf, cd, acdf,$$

$$bcdf, aabcd, ef, ae, be, aabef, ce, acef, bcef, abce, de,$$

$$adef, bdef, abde, cdef, acde, bcde, abcdef.$$

(b) The procedure for a 2^n factorial experiment is the same as for the 2^6 case. First write out the 2^{n-1} treatment conditions for the first $n-1$ letters. Then, append the n-th letter to these 2^{n-1} treatment combinations as required to obtain the same treatments as in the half replicate block being used.

13.45 We are given that

$$[A] = (a) - (b) - (c) + (abc).$$

and, by Exercises 13.16 and 13.17 ,

$$(a) = r[\mu - \alpha_0 + \beta_0 + \gamma_0 - (\alpha\beta)_{00} - (\alpha\gamma)_{00}$$

$$+(\beta\gamma)_{00} - (\alpha\beta\gamma)_{000}] + \sum_{l=1}^{r} \epsilon_{100l}$$

$$(b) = r[\mu + \alpha_0 - \beta_0 + \gamma_0 - (\alpha\beta)_{00} + (\alpha\gamma)_{00}$$

$$-(\beta\gamma)_{00} - (\alpha\beta\gamma)_{000}] + \sum_{l=1}^{r} \epsilon_{010l}$$

$$(c) = r[\mu + \alpha_0 + \beta_0 - \gamma_0 + (\alpha\beta)_{00} - (\alpha\gamma)_{00}$$

$$-(\beta\gamma)_{00} - (\alpha\beta\gamma)_{000}] + \sum_{l=1}^{r} \epsilon_{001l}$$

$$(abc) = r[\mu - \alpha_0 - \beta_0 - \gamma_0 + (\alpha\beta)_{00} + (\alpha\gamma)_{00}$$

$$+(\beta\gamma)_{00} - (\alpha\beta\gamma)_{000}) + \sum_{l=1}^{r} \epsilon_{111l}$$

Thus,

$$[A] = r[-4\alpha_0 + 4(\beta\gamma)_{00}] + \sum_{l=1}^{r} (\epsilon_{100l} - \epsilon_{010l} - \epsilon_{001l} + \epsilon_{111l})$$

$$= -4r[\alpha_0 - (\beta\gamma)_{00}] + \epsilon_A$$

where

$$\epsilon_A = \sum_{l=1}^{r} (\epsilon_{100l} - \epsilon_{010l} - \epsilon_{001l} + \epsilon_{111l})$$

Since the expected values of $\epsilon_{100l}, \epsilon_{010l}$, ϵ_{001l} and ϵ_{111l} are zero, the expected value of ϵ_A is zero.

13.47 First, we arrange the data in modified standard order and calculate the following table:

Experimental conditions	Rep. 1	Rep 2.	Difference	$(y_1 - y_2)^2/2$
1	39.0	43.2	−4.2	8.82
ad	42.0			
bd	54.9			
ab	40.9	40.3	0.6	0.18
cd	43.1			
ac	29.3			
bc	34.8	48.2	−13.4	89.78
abcd	41.4	49.5	−8.1	32.805

The error sum of squares, obtained by summing the last column of the table, is

$$8.82 + 0.18 + 89.78 + 32.805 = 131.585$$

There are 4 degrees of freedom so the mean squared error is

$$\frac{131.585}{4} = 32.896$$

We use Yates methods to calculate effect totals. Similar to Exercise 13.38, we put the appended letter in parentheses to keep in mind that it is nothing to do with the effect totals calculated.

Yates method for Exercise 13.47

Exp. con.	Tr. total	1	2	3	Sum of squares
1	39.0	81.0	176.8	325.4	13235.645
$a(d)$	42.0	95.8	148.6	−18.2	41.405
$b(d)$	54.9	72.4	−11.0	18.6	43.245
ab	40.9	76.2	−7.2	3.4	1.445
$c(d)$	43.1	3.0	14.8	−28.2	99.405
ac	29.3	−14.0	3.8	3.8	1.805
bc	34.8	−13.8	−17.0	−11.0	15.125
$abc(d)$	41.4	6.6	20.4	37.4	174.845

The analysis-of-variance table is

Source of variation	Degrees of freedom	Sums of squares	Mean square	F
Main effects:				
$A = BCD$	1	41.405	41.405	1.259
$B = ACD$	1	43.245	43.245	1.315
$C = ABD$	1	99.405	99.405	3.022
$D = ABC$	1	174.845	174.845	5.315
$AB = CD$	1	1.445	1.445	0.044
$AC = BD$	1	1.805	1.805	0.055
$BC = AD$	1	15.125	15.125	0.460
Error	4	131.585	32.896	
Total	11	508.860		

The critical value is $F_{.05} = 7.71$ with 1 and 4 degrees of freedom. None of the effects or interactions is significant at the 5 percent level .

13.49 (a) As a two-way classification with 12 treatments and three replicates, the analysis-of-variance table is

Source of variation	Degrees of freedom	Sums of squares	Mean square	F
Replication	2	.0039	.00195	.16
Treatments	11	12.7830	1.16209	97.25
Error	22	.2628	.01195	
Total	35	13.0497		

Since $F_{.05} = 3.44$ with 2 and 22 degrees of freedom, the replication effect is not significant at the .05 level. Since $F_{.05} = 2.26$ with 11 and 22 degrees of freedom, the treatment effect is significant at the .05 level.

(b) The three two-way tables:

		(B) Time(min.) 20	30	Totals
(A) Temp.	350^0F	12.7	15.2	27.9
	400^0F	14.6	16.8	31.4
	Totals	27.3	32.0	59.3

		(C) Tenderizer A	B	C	Totals
(A) Temp.	350^0F	9.4	13.3	5.2	27.9
	400^0F	10.2	14.9	6.3	31.4
	Totals	19.6	28.2	11.5	59.3

		(C) Tenderizer A	B	C	Totals
(B) Time	20	8.7	13.1	5.5	27.3
	30	10.9	15.1	6.0	32.0
	Totals	19.6	28.2	11.5	59.3

From the first table, we have

$$\begin{aligned}
SSA &= 1,764.37/(2\cdot 3\cdot 3) - 97.6803 = .3403\\
SSB &= 1,769.29/(2\cdot 3\cdot 3) - 97.6803 = .6136\\
SSC &= 1,311.65/(2\cdot 2\cdot 3) - 97.6803 = 11.6239\\
SS(AB) &= 887.73/(3\cdot 3) - 97.6803 - SSA - SSB = .0025\\
SS(AC) &= 658.03/(2\cdot 3) - 97.6803 - SSA - SSC = .0272\\
SS(BC) &= 660.37/(2\cdot 3) - 97.6803 - SSB - SSC = .1439\\
SS(ABC) &= SS(Tr) - SSA - SSB - SSC - SS(AB) - SS(AC)\\
&\quad - SS(BC) = .0316
\end{aligned}$$

(c) The analysis-of-variance table is

Source of variation	Degrees of freedom	Sums of squares	Mean square	F
Replication	2	.0039	.00195	.16
Main effects:				
A Temperature	1	.3403	.34030	28.48
B Time	1	.6136	.61360	51.35
C Tenderizer	2	11.6239	5.81195	486.36
Two-factor interactions:				
AB	1	.0025	.00250	.21
AC	2	.0272	.01360	1.14
BC	2	.1439	.07193	6.02
ABC interaction:	2	.0316	.01580	1.32
Error	22	.2628	.01195	
Total	35	13.0497		

Since $F_{.05} = 3.44$ with 2 and 22 degrees of freedom, the tenderizer effect and time-tenderizer interaction are significant at the .05 level. Since $F_{.05} = 4.30$ with 1 and 22 degrees of freedom, the main effects of temperature and of time are significant at the .05 level. No other effects are significant at the .05 level.

13.51 The table of $\bar{y}_{ij.}$ for the two factors are

		B 1	2	3	$\bar{y}_{i..}$
A	1	32	16	18	22
	2	14	26	20	20
	$\bar{y}_{.j.}$	23	21	19	

We also have $a = 2$, $b = 3$, $r = 2$ and

$$\bar{y}_{..1} = 19.3333\ , \qquad \bar{y}_{..2} = 22.6667\ , \qquad \bar{y}_{...} = 21$$

Consequently, in each array and summing, we obtain

$$\begin{aligned}
SST &= \sum_{i=1}^{2}\sum_{j=1}^{3}\sum_{k=1}^{2}(y_{ijk} - \bar{y}_{...})^2 \\
&= 8^2 + 14^2 + \cdots + (-5)^2 + 3^2 \;=\; 548 \\
SSA &= (2)(3)\sum_{i=1}^{2}(\bar{y}_{i..} - \bar{y}_{...})^2 \;=\; 6\left(1^2 + (-1)^2\right) \;=\; 12 \\
SSB &= (2)(2)\sum_{j=1}^{3}(\bar{y}_{.j.} - \bar{y}_{...})^2 \;=\; 4\left(2^2 + 0^2 + (-2)^2\right) \;=\; 32 \\
SSR &= (2)(3)\sum_{k=1}^{2}(\bar{y}_{..k} - \bar{y}_{...})^2 \;=\; 6\left((-1.6667)^2 + 1.6667^2\right) \;=\; 33.3347
\end{aligned}$$

$$\begin{aligned} SS(AB) &= 2\sum_{i=1}^{2}\sum_{j=1}^{3}(\bar{y}_{ij.} - \bar{y}_{i..} - \bar{y}_{.j.} + \bar{y}_{...})^2 \\ &= 2\left(8^2 + (-6)^2 + (-2)^2 + (-8)^2 + 6^2 + 2^2\right) = 416 \\ SSE &= SST - SSA - SSB - SSR - SS(AB) = 54.6653 \end{aligned}$$

The analysis-of-variance table is

Source of variation	Degrees of freedom	Sums of squares	Mean square	F
Replication	1	33.3347	33.3347	3.05
A	1	12.0000	12.0000	1.10
B	2	32.0000	16.0000	1.46
AB interaction:	2	416.0000	208.0000	19.03
Error	5	54.6653	10.9331	
Total	11	548.0000		

Since $F_{.05} = 6.61$ with 1 and 5 degrees of freedom, and $F_{.05} = 5.79$ with 2 and 5 degrees of freedom, only the AB interaction is significant at .05 level.

13.53 The visual summary of the four treatment means is given in Figure 13.5. According to the visual

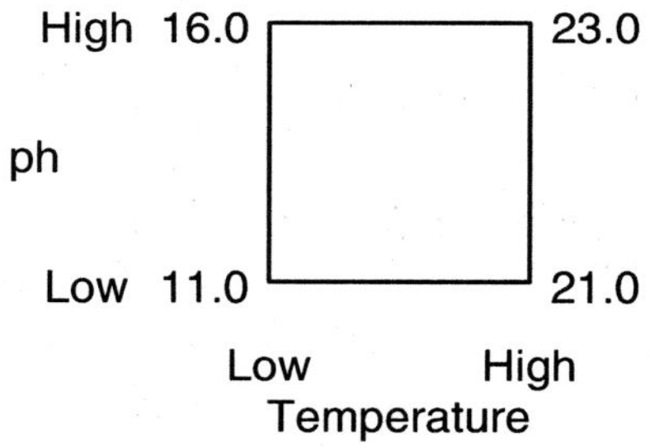

Figure 13.5: Visual summary of the experiment. Exercise 13.52

procedure, we assume that there is no replication effects in the model. The error sum of squares is

$$SSE = \sum_{i=0}^{1}\sum_{j=0}^{1}\sum_{l=0}^{r}(y_{ijl} - \bar{y}_{ij.})^2$$

with $2^2(r-1)$ degrees of freedom. Thus we calculate

	Rep. 1	Rep. 2	Rep. 3	Mean	$y_{ij1}-\bar{y}_{ij.}$	$y_{ij2}-\bar{y}_{ij.}$	$y_{ij3}-\bar{y}_{ij.}$
1	10	14	9	11	−1	3	−2
a	21	19	23	21	0	−2	2
b	17	15	16	16	1	−1	0
ab	20	24	25	23	−3	1	2
Mean	17	18	18.25	17.75			

Squaring the entries in the last three columns and summing, we obtain

$$SSE = ((-1)^2 + 3^2 + (-2)^2 + \cdots + 2^2) = 38$$

with $2^2(r-1) = 4(2) = 8$ degrees of freedom. Therefore, $s^2 = MSE = 38/8 = 4.75$. Since $t_{.025} = 2.306$ for 8 degrees of freedom, the half length of the confidence intervals is

$$t_{.025}\sqrt{\frac{s^2}{r}} = 2.306\sqrt{\frac{4.75}{3}} = 2.90$$

and the confidence intervals are:

Factor A :	8.5 ± 2.9 or 5.6 to 11.4
Factor B :	3.5 ± 2.9 or .6 to 6.4
Interaction AB :	-1.5 ± 2.9 or -4.4 to 1.4

The confidence interval for interaction overs 0 but both those for main effects do not.

13.55 The visual summary of the eight treatment means is given in Figure 13.6. According to visual procedure, we assume that there is no replication effects in the model. The error sum of squares is

$$SSE = \sum_{i=0}^{1}\sum_{j=0}^{1}\sum_{k=0}^{1}\sum_{l=1}^{r}(y_{ijkl} - \bar{y}_{ijk.})^2$$

with $2^3(r-1)$ degrees of freedom. Thus we calculate

	Rep. 1	Rep. 2	Mean	$y_{ijk1}-\bar{y}_{ijk.}$	$y_{ijk2}-\bar{y}_{ijk.}$
1	41.8	42.2	42.0	−.2	.2
a	44.5	43.9	44.2	.3	−.3
b	56.5	56.3	56.4	.1	−.1
ab	57.3	56.5	56.9	.4	−.4
c	43.4	42.7	43.05	.35	−.35
ac	42.5	43.1	42.8	−.3	.3
bc	56.5	55.3	55.9	.6	−.6
abc	56.5	55.6	56.05	.45	−.45

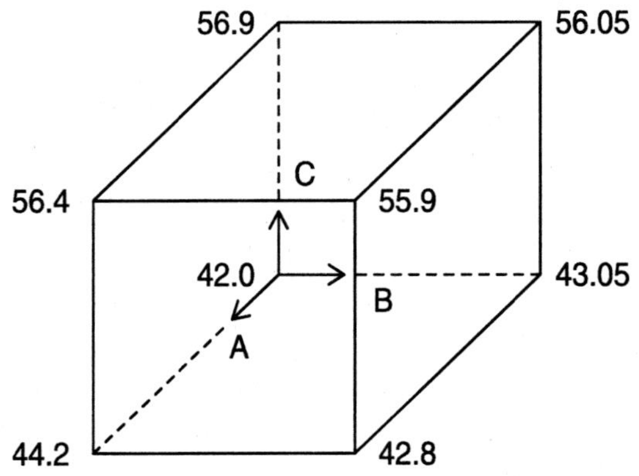

Figure 13.6: Visual summary of the experiment. Exercise 13.55

Note that $y_{ijk1} - \bar{y}_{ijk.} = -(y_{ijk2} - \bar{y}_{ijk.})$. Squaring the entries in the last two columns and summing, or taking twice the sum of squares for one column, we obtain

$$SSE = 2(\ (-.2)^2 + (.3)^2 + \cdots + (.45)^2) = 2.150$$

with $2^3(2-1) = 8$ degrees of freedom. Therefore, $s^2 = MSE = 2.150/8 = .26875$. Since $t_{.025} = 2.306$ for 8 degrees of freedom, the half length of the confidence intervals is

$$t_{.025}\sqrt{\frac{s^2}{2r}} = 2.306\sqrt{\frac{.26875}{4}} = .5977$$

The estimated effects are:

Factor A :

$$\bar{y}_{1...} - \bar{y}_{0...} = 49.9875 - 49.3375 = .6500$$

Factor B :

$$\bar{y}_{.1..} - \bar{y}_{.0..} = 56.3125 - 43.0125 = 13.3000$$

Factor C :

$$\bar{y}_{..1.} - \bar{y}_{..0.} = 49.4500 - 49.875 = -.4250$$

AB :

$$\frac{1}{2}(\bar{y}_{11..} - \bar{y}_{10..} - \bar{y}_{01..} + \bar{y}_{00..}) = \frac{1}{2}(56.4750 - 43.5000 - 56.1500 + 42.5250) = -.3250$$

AC :

$$\frac{1}{2}(\bar{y}_{1.1.} - \bar{y}_{1.0.} - \bar{y}_{0.1.} + \bar{y}_{0.0.}) = \frac{1}{2}(49.4250 - 50.5500 - 49.4750 + 49.2000) = -.7000$$

BC :

$$\frac{1}{2}(\bar{y}_{.11.} - \bar{y}_{.10.} - \bar{y}_{.01.} + \bar{y}_{.00.}) = \frac{1}{2}(55.9750 - 56.6500 - 42.9250 + 43.1000) = -.2500$$

ABC :

$$\frac{1}{4}(\bar{y}_{111.} - \bar{y}_{101.} - \bar{y}_{011.} + \bar{y}_{001.}) - \frac{1}{4}(\bar{y}_{110.} - \bar{y}_{100.} - \bar{y}_{010.} + \bar{y}_{000.})$$

$$= \frac{1}{4}(56.0500 - 42.8000 - 55.9000 + 43.0500) - \frac{1}{4}(56.9000 - 44.2000 - 56.4000 + 42.0000) = .5250$$

so the confidence intervals are:

A :	$6.5000 \pm .5977$ or 5.90 to 7.10
B :	$13.3000 \pm .5977$ or 12.70 to 13.90
C :	$-.4250 \pm .5977$ or -1.02 to .017
AB :	$.3250 \pm .5977$ or $-.27$ to .92
AC :	$-.7000 \pm .5977$ or -1.30 to $-.10$
BC :	$-.2500 \pm .5977$ or $-.85$ to .35
ABC :	$.5250 \pm .5977$ or $-.07$ to 1.12

The confidence interval for the AC interaction does not cover 0 so those two factors must be considered jointly. Factor B has the largest non-zero effect. Using the larger shot size increases tensile strength from 12.70 to 13.90 coded units.

13.57 Using the table in Exercise 13.53, we have

$$\begin{aligned} SST &= \sum_{i=0}^{1}\sum_{j=0}^{1}\sum_{l=1}^{r}(y_{ijl} - \bar{y}_{...})^2 \\ &= (-7.75)^2 + (-3.75)^2 + \cdots + 6.75^2 + 7.75^2 = 298.25 \\ SS(Tr) &= r\sum_{i=0}^{1}\sum_{j=0}^{1}(\bar{y}_{ij.} - \bar{y}_{...})^2 \\ &= 3((-6.75)^2 + (-1.75)^2 + (3.25)^2 + (5.25)^2) = 260.25 \end{aligned}$$

We use the Yates method to calculate the sum of squares of factors and interactions.

Yates method for Exercise 13.57				
Exp. con.	Tr. total	1	2	Sum of squares
1	33	96	213	3,780.75
a	63	117	51	216.75
b	48	30	21	36.75
ab	69	21	−9	6.75

The analysis-of-variance table is

Source of variation	Degrees of freedom	Sums of squares	Mean square	F
A	1	216.75	216.75	45.63
B	1	36.75	36.75	7.74
AB	1	6.75	6.75	1.42
Error	8	38.00	4.75	
Total	11	298.25		

The critical values are $F_{.05} = 5.32$ and $F_{.01} = 11.26$, each with 1 and 8 degrees of freedom. The effect for A is significant at the .01 level. The effect for B is significant at the .05 level.

13.59 The largest number of blocks in which one can perform a 2^6 factorial experiment without confounding any main effects is $2^5 = 32$. It is clear that it is impossible to divide the experiment into 2^6 without confounding everything. If we confound on AB , BC , CD , DE, and EF we will have the desired blocks because the generalized interactions between these blocks do not include any main effects.

13.61 (a) The four blocks are

Block 1: a, abc, bd, cd, be, ce, ade, $abcde$

Block 2: b, c, ad, $abcd$, ae, $abce$, bde, cde

Block 3: ab, ac, d, bcd, e, bce, $abde$, $acde$

Block 4: 1, bc, abd, acd, abe, ace, de, $bcde$

(b) The block totals are(see Exercise 13.12):

Block 1: Rep 1: $4+2+3+10+11+4+15+16=65$

Rep 2: $9+4+7+6+5+8+9+11=59$

Block 2: Rep 1: $2+2+8+11+7+17+4+17=68$

Rep 2: $8+5+2+15+4+23+11+11=79$

Block 3: Rep 1: $15+11+0+6+3+4+10+5=54$

Rep 2: $7+6+3+14+7+8+6+10=61$

Block 4: Rep 1: $3+4+5+6+10+19+7+14=68$

Rep 2: $1+1+12+1+17+13+4+9=58$

Thus, $C = 512^2/64 = 4{,}096$ and the $SS(Bl) = 33{,}196/8 - 4{,}096 = 53.5$.

The analysis-of-variance table is

Source of variation	Degrees of freedom	Sums of squares 10^{-6}	Mean square 10^{-6}	F
Blocks	7	53.5000	7.6430	.5900
Main Effects:				
A	1	182.2500	182.2500	12.70
B	1	81.0000	81.0000	5.64
C	1	85.5625	85.5625	5.96
D	1	9.0000	9.0000	.63
E	1	248.0625	248.0625	17.28
Unconfounded Interactions:				
AB	1	16.0000	16.0000	1.17
AC	1	3.0625	3.0625	.21
AD	1	90.2500	90.2500	6.29
AE	1	7.5625	7.5625	.53
BC	1	7.5625	7.5625	.53
BD	1	1.0000	1.0000	.07
BE	1	.5625	.5625	.04
CD	1	22.5625	22.5625	1.57
CE	1	30.2500	30.2500	2.11
DE	1	10.5625	10.5625	.74
ABD	1	1.0000	1.0000	.07
ABE	1	3.0625	3.0625	.21
ACD	1	68.0625	68.0625	4.74
ACE	1	25.0000	25.0000	1.74
BCD	1	60.0625	60.0625	4.18
BCE	1	4.0000	4.0000	.28
BDE	1	60.0625	60.0625	4.18
CDE	1	36.0000	36.0000	2.51
$ABCD$	1	52.5625	52.5625	3.66
$ABCE$	1	9.0000	9.0000	.63
$ABDE$	1	18.0625	18.0625	1.26
$ACDE$	1	42.2500	42.2500	2.94
$ABCDE$	1	.2500	.2500	.02
Intrablock Error	28	401.8750	14.3527	
Total	63	1630.0000		

The critical value is $F_{.05} = 4.20$ with 1 and 28 degrees of freedom. The effects for A, B, C, E and AD, ACD interactions are significant at the .05 level. The result is similar to that of Exercise 13.12.

13.63 We choose the experimental conditions which have an even number of letters in common with the defining contrast $ABCDE$ and put them in the generalized standard order. The defining relation for the half replicate is $I = -ABCDE$. The table of calculations using the Yates method is given in the first table. We put "e" in parentheses to remind us that "e" has nothing to do with the analysis.

Yates method for Exercise 13.63							
Exp. con.	Tr. total	1	2	3	4	Id.	Sum of squares
1	20	47	82	143	282	$I = -ABCDE$	4,970.25
$a(e)$	27	35	61	139	92	$A = -BCDE$	529.00
$b(e)$	9	38	76	43	−52	$B = -ACDE$	169.00
ab	26	23	63	49	26	$AB = -CDE$	42.25
$c(e)$	14	43	24	−27	−34	$C = -ABDE$	72.25
ac	24	33	19	−25	−4	$AC = -BDE$	1.00
bc	7	39	24	9	−8	$BC = -ADE$	4.00
$abc(e)$	16	24	25	17	−14	$ABC = -DE$	12.25
$d(e)$	18	7	−12	−21	−4	$D = -ABCE$	1.00
ad	25	17	−15	−13	6	$AD = -BCE$	2.25
bd	8	10	−10	−5	2	$BD = -ACE$	.25
$abd(e)$	25	9	−15	1	8	$ABD = -CE$	4.00
cd	15	7	10	−3	8	$CD = -ABE$	4.00
$acd(e)$	24	17	−1	−5	6	$ACD = -BE$	2.25
$bcd(e)$	4	9	10	−11	−2	$BCD = -AE$	.25
$abcd$	20	16	7	−3	8	$ABCD = -E$	4.00

We will assume that the two-way interactions are due to random error. The analysis-of-variance table is

Source of variation	Degrees of freedom	Sums of squares	Mean square	F
Confounded main effects:				
A	1	529.00	529.00	72.97
B	1	169.00	169.00	23.31
C	1	72.25	72.25	9.97
D	1	1.00	1.00	.14
E	1	4.00	4.00	.55
Error	10	72.50	7.25	
Total	15	847.75		

The critical value is $F_{.05} = 4.96$ with 1 and 10 degrees of freedom. The main effects for A, B and C are significant at the .05 level.

Chapter 14

NONPARAMETRIC TESTS

14.1 1. *Null hypothesis:* $\tilde{\mu} = 0.55$ $(p = \frac{1}{2})$

Alternative hypothesis: $\tilde{\mu} \neq 0.55$ $(p > \frac{1}{2})$

2. *Level of significance:* $\alpha = 0.05$.

3. *Criterion:* The criterion may be based on the number of plus signs or the number of minus signs. Using the number of plus signs, denoted by x, reject the null hypothesis if either the probability of getting x or more plus signs or the probability of getting x or fewer plus signs is less than or equal to 0.05/2.

4. *Calculations:* The signs of the differences between the observations and $\tilde{\mu} = .55$ are:

$$+ \; + \; - \; 0 \; + \; 0 \; - \; + \; + \; - \; + \; + \; - \; + \; + \; - \; + \; -$$

Ignoring the two 0's, for the cases where the values equal 0.55, we have x =10 and the effective sample size is n =16. From the binomial distribution, Table 1, the probability of $X \geq 10$ with $p = 0.5$ is $1 - .7728 = .2272$. The probability of $X \leq 10$ is .8949.

5. *Decision:* Since both of the probabilities are greater than $0.05/2 = .025$, we fail to reject the null hypothesis $\tilde{\mu} = 0.55$, at the level $\alpha = .05$.

14.3 1. *Null hypothesis:* $\tilde{\mu} = 6.5$ $(p = \frac{1}{2})$

Alternative hypothesis: $\tilde{\mu} < 6.5$ $(p < \frac{1}{2})$

2. *Level of significance:* $\alpha = 0.01$.

3. *Criterion:* The criterion may be based on the number of plus signs or the number of minus signs. Using the number of plus signs, denoted by x, reject the null hypothesis if the probability of getting x or fewer positives is less than or equal to 0.01. Using the normal approximation, we

reject H_0 if

$$z = \frac{x + .5 - n/2}{\sqrt{n(1/2)(1/2)}} < -2.327$$

4. *Calculations:* The signs of the differences between the observations and $\tilde{\mu} = 150$ are:

$$- \; - \; - \; - \; - \; - \; - \; - \; - \; - \; - \; - \; - \; - \; - \; - \; - \; - \; - \; -$$

$$+ \; - \; - \; - \; + \; - \; + \; + \; - \; - \; - \; + \; - \; + \; - \; - \; + \; + \; - \; -$$

$$+ \; + \; - \; + \; - \; - \; - \; - \; + \; - \; - \; + \; + \; + \; + \; - \; - \; - \; - \; -$$

$$- \; - \; - \; - \; - \; - \; + \; - \; + \; - \; - \; - \; - \; - \; + \; - \; + \; - \; + \; +$$

The sample size is $n = 80$ and we have $x = 22$ positives. Using the normal approximation to the binomial with $p = 0.5$,

$$z = \frac{22.5 - 80/2}{\sqrt{80(1/2)(1 - 1/2)}} = -3.91$$

5. *Decision:* Since $z = -3.91$, we reject the null hypothesis in favor of $\tilde{\mu} < 6.5$, at the level $\alpha = .01$.

14.5 1. *Null hypothesis:* Populations are identical.

Alternative hypothesis: Populations are not identical.

2. *Level of significance:* $\alpha = 0.01$.

3. *Criterion:* We reject the null hypothesis if U_1 is too small or too large. That is, we reject H_0 if

$$Z = \frac{U_1 - \mu_{U_1}}{\sigma_{U_1}} < -2.575 \quad or \quad Z > 2.575.$$

4. *Calculations:* The ranks for the first sample are

$$16, 21, 5, 14.5, 13, 12, 18, 22, 20, 23, 17, 19$$

The sum of ranks of the first sample is

$$W_1 = 200.5$$

Thus,

$$U_1 = W_1 - \frac{n_1(n_1+1)}{2} = 200.5 - \frac{12 \cdot 13}{2} = 122.5.$$

Under the null hypothesis, the mean and variance of the U_1 statistic are

$$\mu_{U_1} = \frac{n_1 \cdot n_2}{2} = \frac{12 \cdot 12}{2} = 72.$$

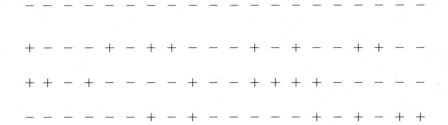

and

$$\sigma^2_{U_1} = \frac{n_1 \cdot n_2(n_1 + n_2 + 1)}{12} = \frac{12 \cdot 12(12 + 12 + 1)}{12} = 300.$$

Thus, the Z statistic is

$$Z = \frac{122.5 - 72}{\sqrt{300}} = 2.916$$

5. *Decision:* We reject the null hypothesis in favor of the alternative that the populations are not identical, at the .01 level of significance.

14.7 1. *Null hypothesis:* Populations are identical.

Alternative hypothesis: The first population is stochastically larger than the second.

2. *Level of significance:* $\alpha = 0.05$.

3. *Criterion:* We reject the null hypothesis if U_1 is too large. That is, we reject H_0 if

$$Z = \frac{U_1 - \mu_{U_1}}{\sigma_{U_1}} > 1.645$$

4. *Calculations:* The ranks for the first sample are

$$3, 20, 32, 25, 11.5, 14, 24, 27.5, 17.5, 21, 1, 22, 29, 10, 19, 30.5$$

The sum of ranks of the first sample is

$$W_1 = 307$$

Thus,

$$U_1 = W_1 - \frac{n_1(n_1 + 1)}{2} = 307 - \frac{16 \cdot 17}{2} = 171.$$

Under the null hypothesis, the mean and variance of the U_1 statistic are

$$\mu_{U_1} = \frac{n_1 \cdot n_2}{2} = \frac{16 \cdot 16}{2} = 128$$

and

$$\sigma^2_{U_1} = \frac{n_1 \cdot n_2(n_1 + n_2 + 1)}{12} = \frac{16 \cdot 16(16 + 16 + 1)}{12} = 704.$$

Thus, the Z statistic is

$$Z = \frac{171 - 120}{\sqrt{704}} = 1.621$$

5. *Decision:* We cannot reject the null hypothesis at the .05 level of significance. In other words , we cannot conclude that strength of material 1 is stochastically larger than that of material 2.

14.9 1. *Null hypothesis:* The populations are identical.

Alternative hypothesis: Populations are not all equal.

2. *Level of significance:* $\alpha = 0.05$.

3. *Criterion:* We reject the null hypothesis if $H > 9.488$ the value of $\chi^2_{.05}$ for 4 degrees of freedom.

4. *Calculations:* The sums of ranks for the samples are

$$\begin{aligned}
R_1 &= 20.5 + 22 + 23.5 + 23.5 + 26.5 + 31 = 147.0 \\
R_2 &= 5 + 12 + 13.5 + 13.5 + 15 + 17 + 20.5 = 96.5 \\
R_3 &= 1 + 2 + 3 + 7 + 7 + 10.5 = 30.5 \\
R_4 &= 19 + 25 + 26.5 + 28 + 29 + 30 + 32 + 33 = 222.5 \\
R_5 &= 4 + 7 + 9 + 10.5 + 16 + 18 = 64.5
\end{aligned}$$

Thus,

$$H = \frac{12}{30 \cdot 31}\left[\frac{(147)^2}{6} + \frac{(96.5)^2}{7} + \frac{(30.5)^2}{6} + \frac{(222.5)^2}{8} + \frac{(64.5)^2}{6}\right] - 3 \cdot 34 = 26.01.$$

5. *Decision:* We reject the null hypothesis at the .05 level of significance. The P-value is much smaller than .005.

14.11 1. *Null hypothesis:* Arrangement is random.
Alternative hypothesis: Arrangement is not random.

2. *Level of significance:* $\alpha = 0.05$.

3. *Criterion:* We reject the null hypothesis if

$$Z = \frac{u - \mu_u}{\sigma_u} < -1.96 \quad or \quad Z > 1.96.$$

where u is the total number of runs and

$$\mu_u = \frac{2n_1 n_2}{n_1 + n_2} + 1$$

and

$$\sigma_u^2 = \frac{2n_1 n_2(2n_1 n_2 - n_1 - n_2)}{(n_1 + n_2)^2(n_1 + n_2 - 1)}$$

4. *Calculations:* The runs in the data are underlined.

$\underline{LL}\ \underline{O}\ \underline{LLLL}\ \underline{OO}\ \underline{LLLL}\ \underline{O}\ \underline{L}\ \underline{OO}\ \underline{LLLL}\ \underline{O}\ \underline{L}\ \underline{OO}\ \underline{LLLLL}$

$\underline{O}\ \underline{LLL}\ \underline{OL}\ \underline{O}\ \underline{LLLL}\ \underline{OO}\ \underline{L}\ \underline{OOOO}\ \underline{LLLL}\ \underline{O}\ \underline{L}\ \underline{OO}\ \underline{LLL}\ \underline{O}$

There are $u = 28$ runs, 22 O's and 38 L's. Under the null hypothesis, the mean and standard deviation are

$$\begin{aligned}\mu_u &= \frac{2n_1n_2}{n_1+n_2}+1 = \frac{2\cdot 22\cdot 38}{22+38}+1 = 28.87\\ \sigma_u &= \sqrt{\frac{2n_1n_2(2n_1n_2-n_1-n_2)}{(n_1+n_2)^2(n_1+n_2-1)}}\\ &= \sqrt{\frac{2\cdot 22\cdot 38(2\cdot 22\cdot 38-22-38)}{(22+38)^2(22+38-1)}} = 3.562\end{aligned}$$

Thus, the Z statistic is

$$Z = \frac{28-28.87}{3.562} = -.244$$

5. *Decision:* We cannot reject the null hypothesis at the .05 level of significance.

14.13 1. *Null hypothesis:* The arrangement of sample values is random.

Alternative hypothesis: The arrangement is not random.

2. *Level of significance:* $\alpha = 0.05$.

3. *Criterion:* We reject the null hypothesis if

$$Z = \frac{u - \mu_u}{\sigma_u} < -1.96 \quad or \quad Z > 1.96.$$

where u is the total number of runs above and below the median.

4. *Calculations:* The median of the data is 36. Using a for observations above the median and b for those below, the data are:

$$\underline{bbbbbb}\ \underline{a}\ \underline{bbbbbbb}\ \underline{aa}\ \underline{bbbbbbb}\ \underline{aaa}\ \underline{bbb}\ \underline{aa}\ \underline{b}\ \underline{aaaaa}\ \underline{b}\ \underline{aaaaaaaaaaa}$$

There are $u = 12$ runs, 25 a's and 25 b's. In this case

$$\mu_u = \frac{2\cdot 25\cdot 25}{25+25}+1 = 26$$

and

$$\sigma_u = \sqrt{\frac{2\cdot 25\cdot 25(2\cdot 25\cdot 25-25-25}{25+25)^2(25+25-1)}} = 3.448$$

so

$$Z = \frac{12-26}{3.449} = -4.005$$

5. *Decision:* We reject the null hypothesis. at the .05 level of significance. The P- value is .00006.

14.15 The null hypothesis is that the data are from an exponential distribution with mean 10. This is also the Weibull distribution with shape = 1 and scale = 10. The alternative is that the data are not from this exponential distribution. To test the null hypothesis against the alternative, at the 5 percent level, use the statistic D. According to the computer output, we cannot reject the null hypothesis. The $P-$value is greater than 0.10.

```
x=c(1.5,10.3,3.6,13.4,18.4,7.7,24.3,10.7,8.4,15.4,
        + 4.9,2.8,7.9,11.9,12.0,16.2,6.8,14.7)
ks.test(x,"pweibull",shape=1,scale=10)

        One-sample Kolmogorov-Smirnov test

data:  x
D = 0.2712, p-value = 0.1166
alternative hypothesis: two-sided
```

The value $D = .2712$ is the maximum difference between the empirical and postulated distributions illustrated in Figure 14.1.

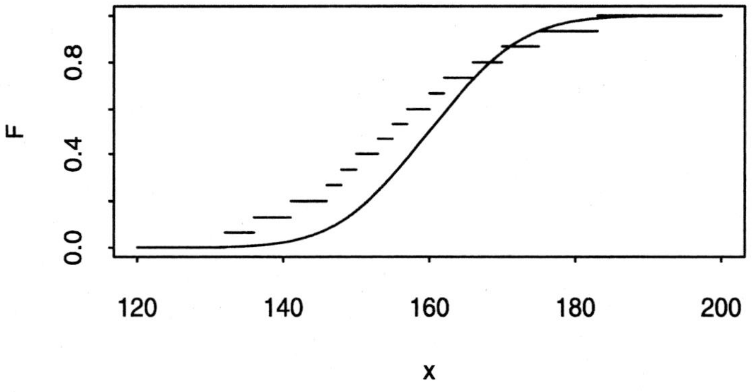

Figure 14.1: Empirical and null distributions. Exercise 14.14.

14.17 1. *Null hypothesis:* Populations are identical.

Alternative hypothesis: The Method B population is stochastically larger than the Method A population.

2. *Level of significance:* $\alpha = 0.05$.

3. *Criterion:* We reject the null hypothesis if U_1 is too small. That is, we reject H_0 if

$$Z = \frac{U_1 - \mu_{U_1}}{\sigma_{U_1}} < -1.645.$$

4. *Calculations:* The ranks for the first sample are

$$1.5,\ 3,\ 4.5,\ 4.5,\ 6.5,\ 9.5,\ 9.5,\ 12.5,\ 14,\ 15.5$$

The sum of ranks of the first sample is

$$W_1 = 1.5 + 3 + 4.5 + 4.5 + 6.5 + 9.5 + 9.5 + 12.5 + 14 + 15.5 = 81$$

Thus ,

$$U_1 = W_1 - \frac{n_1(n_1+1)}{2} = 81 - \frac{10 \cdot 11}{2} = 26.$$

Under the null hypothesis, the mean and variance of the U_1 statistic are

$$\mu_{U_1} = \frac{n_1 \cdot n_2}{2} = \frac{10 \cdot 10}{2} = 50.$$

and

$$\sigma^2_{U_1} = \frac{n_1 \cdot n_2(n_1+n_2+1)}{12} = \frac{10 \cdot 10(10+10+1)}{12} = 175.$$

Thus, the Z statistic is

$$Z = \frac{26-50}{\sqrt{175}} = -1.814$$

5. *Decision:* We reject the null hypothesis at the .05 level of significance and conclude that the Method B population is stochastically larger than the Method A population.

14.19

1. *Null hypothesis:* The three populations are identical.
   Alternative hypothesis: Populations are not all equal.
2. *Level of significance:* $\alpha = 0.05$.
3. *Criterion:* We reject the null hypothesis if $H > 5.991$ the value of $\chi^2_{.05}$ for 2 degrees of freedom.
4. *Calculations:* The sums of ranks for the samples are

$$R_1 = 1.5 + 5 + 7.5 + 10.5 + 12 + 13 + 15.5 + 18 + 25 + 28 = 136$$

$$R_2 = 3 + 5 + 7.5 + 9 + 10.5 + 20 + 21 + 22.5 + 28 + 30 = 156.5$$

$$R_3 = 1.5 + 5 + 14 + 15.5 + 18 + 18 + 22.5 + 25 + 25 + 28 = 172.5$$

Thus ,

$$H = \frac{12}{30 \cdot 31}\left[\frac{(136)^2}{10} + \frac{(156.5)^2}{10} + \frac{(172.5)^2}{10}\right] - 3 \cdot 31 = 0.904.$$

5. *Decision:* We cannot reject the null hypothesis at the .05 level of significance.

14.21 1. *Null hypothesis:* The arrangement of sample values is random.

Alternative hypothesis: There is a trend in the sample.

2. *Level of significance:* $\alpha = 0.05$.

3. *Criterion:* We reject the null hypothesis if

$$Z = \frac{u - \mu_u}{\sigma_u} < -1.645.$$

where u is the total number of runs above and below the median.

4. *Calculations:* The median of the data is 138. Using a for observations above the median and b for those below, the data are:

$$\underline{bbbbb}\ \underline{aa}\ \underline{bb}\ \underline{aaaa}\ \underline{b}\ \underline{a}\ \underline{bbb}\ \underline{aaaa}\ \underline{bbb}\ \underline{a}\ \underline{bb}\ \underline{s}\ \underline{aaaa}$$

Ignoring the tie (s), there are $u = 12$ runs, 16 a's and 16 b's. In this case

$$\mu_u = \frac{2 \cdot 16 \cdot 16}{16 + 16} + 1 = 17$$

and

$$\sigma_u = \sqrt{\frac{2 \cdot 16 \cdot 16(2 \cdot 16 \cdot 16 - 16 - 16}{16 + 16)^2(16 + 16 - 1)}} = 2.782$$

so

$$Z = \frac{12 - 17}{2.782} = -1.797$$

5. *Decision:* We reject the null hypothesis at the .05 level of significance and conclude that there is a trend over time.

14.23 1. *Null hypothesis:* Populations are identical.

Alternative hypothesis: The Heat 1 population is stochastically larger than the Heat 2 population.

2. *Level of significance:* $\alpha = 0.033$.

3. *Criterion:* We reject the null hypothesis if U_1 is too large. Since the distribution of U_1 is symmetric about $n_1 n_2 / 2 = 10.5$, $.033 = P(U_1 \leq 2) = P(U_1 \geq 19)$ and we reject H_0 if $U_1 \geq 19$.

4. *Calculations:* The sum of ranks of the first sample is

$$W_1 = 6 + 9 + 10 = 25$$

Thus,

$$U_1 = W_1 - \frac{n_1(n_1 + 1)}{2} = 25 - \frac{3 \cdot 4}{2} = 19.$$

5. *Decision:* We reject the null hypothesis at the .033 level of significance and conclude that the Heat 1 population is stochastically larger than the Heat 2 population.

14.25 1. *Null hypothesis:* The arrangement of sample values is random.
Alternative hypothesis: The arrangement is not random.

2. *Level of significance:* $\alpha = 0.05$.

3. *Criterion:* We reject the null hypothesis if

$$Z = \frac{u - \mu_u}{\sigma_u} < -1.96 \quad \text{or} \quad Z > 1.96.$$

where u is the total number of runs above and below the median.

4. *Calculations:* The median of the differences is 2. Using a for observations above the median and b for those below, the data are:

$$\underline{s}\ \underline{bbbbb}\ \underline{a}\ \underline{b}\ \underline{s}\ \underline{aaaaaaa}\ \underline{s}\ \underline{bbb}$$

Ignoring the ties (s), there are $u = 5$ runs, $n_1 = 8$ a's and $n_2 = 9$ b's. In this case

$$\mu_u = \frac{2 \cdot 8 \cdot 9}{8 + 9} + 1 = 9.471$$

and

$$\sigma_u = \sqrt{\frac{2 \cdot 8 \cdot 9(2 \cdot 8 \cdot 9 - 8 - 9)}{(8+9)^2(8+9-1)}} = 1.989$$

so

$$Z = \frac{6 - 9.471}{1.989} = -2.243$$

5. *Decision:* We reject the null hypothesis at the .05 level of significance. There are two few runs suggesting that adjacent positions are positively correlated.

Chapter 15

THE STATISTICAL CONTENT OF QUALITY IMPROVEMENT PROGRAMS

15.1 (a) The central line for the $\bar{x}$ chart is given by $y = \mu = 0.150$. The lower control limit is given by

$$LCL = 0.150 - \frac{3}{\sqrt{5}}(0.002) = 0.147$$

and the upper control limit is given by

$$UCLy = 0.150 + \frac{3}{\sqrt{5}}(0.002) = 0.153.$$

(b) The central line for the R chart is given by $y = d_2\sigma = 2.326(0.002) = 0.005$. The lower control limit is given by $LCL = D_1\sigma = 0(0.002) = 0$ and the upper control limit is given by $UCL = D_2\sigma = 4.918(0.002) = 0.010$.

(c) The control charts are given in the Figures 15.1 and 15.2. For the $\bar{x}$ chart, points at 8, 16, and 17 are outside the limits. For the R chart , all points are within the limits.

15.3 We first find $\bar{\bar{x}} = 48.10$ and $\bar{R} = 2.95$.

(a) The central line for the $\bar{x}$ chart is $\bar{\bar{x}} = 48.10$, the $LCL = 45.95$ and the $UCL = 50.25$.

(b) The center line for the R chart is $\bar{R} = 2.95$, the $LCL = 0$ and the $UCL = 6.73$.

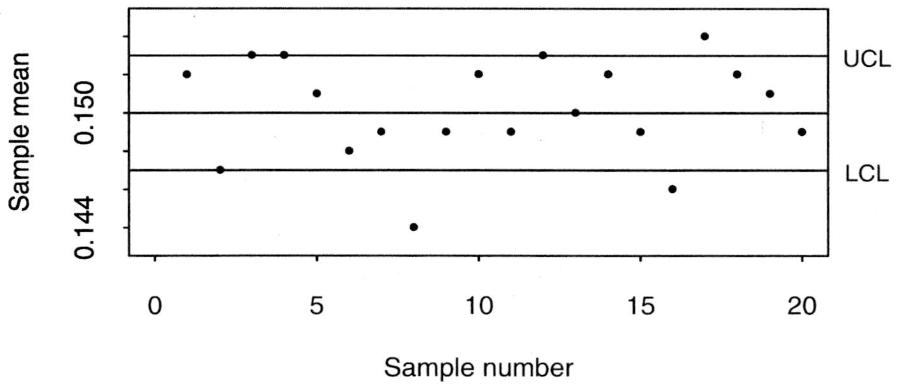

Figure 15.1: Control chart for sample means for Exercise 15.1.

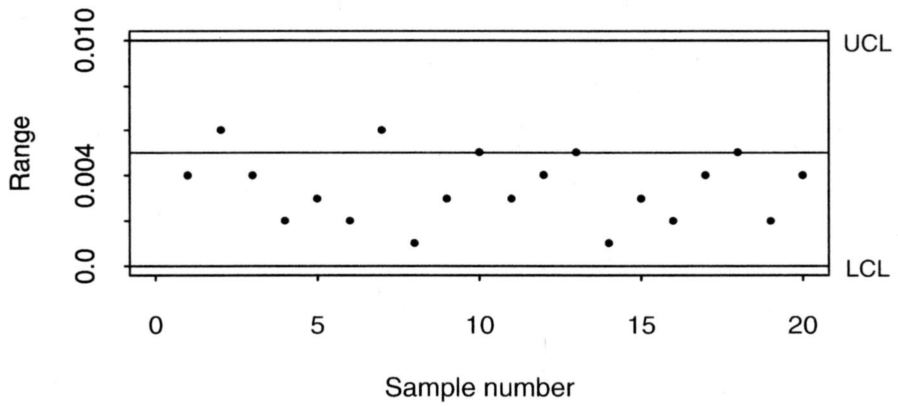

Figure 15.2: Control chart for sample ranges for Exercise 15.1.

(c) The control charts are given in the Figures 15.3 and 15.4. For the $\bar{x}$ chart, there are points outside the limits at 5, 7, 8, 9, 11, 12, 13, 15, 16, 17, 18, 21, 24, and 25. For the R chart, all points are within the limits.

(d) Their are 8 runs with 11 points below and 13 above.

$$\mu_u = 12.92, \sigma_u = 2.38 \text{ so } z = -2.07$$

This is significant at the $\alpha = 0.025$ level ($z_\alpha = 1.96$).

(e) No, since the process in not in control.

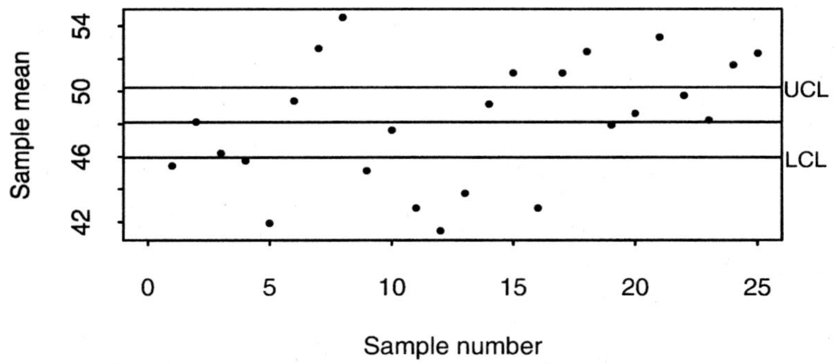

Figure 15.3: Control chart for sample means for Exercise 15.3.

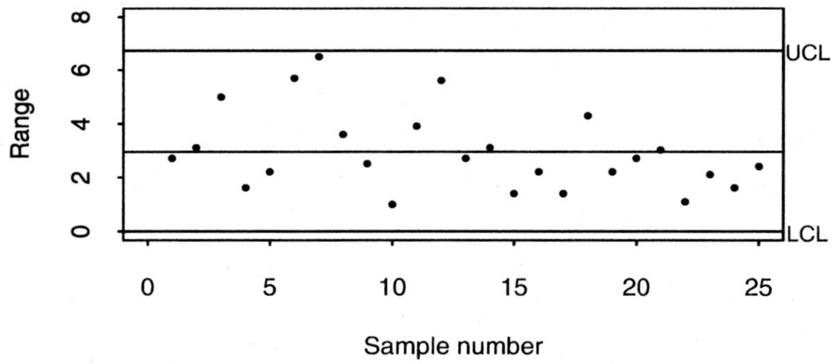

Figure 15.4: Control chart for sample ranges for Exercise 15.3.

15.5 (a) We calculate $\bar{\bar{x}} = 21.7$, $\bar{s} = 1.455$. Thus, for the $\bar{x}$ chart, the center line is 21.7 and

$$LCL = \bar{x} - A_1 \bar{s} = 21.7 - 2.394(1.455) = 18.22$$

$$UCL = \bar{x} + A_1 \bar{s} = 21.7 + 2.394(1.455) = 25.18$$

The σ chart has center line $c_2 \bar{s} = .72356\ (\ 1.455) = 1.053$

$$LCL = B_3 \bar{s} = 0(1.445) = 0$$

$$UCL = B_4 \bar{s} = 2.568(1.445) = 3.736$$

The control charts are given in Figures 15.5 and 15.6. All of the sample means are within the control limits. However, at first, there is a run of nine points below the central line. Only one sample standard deviation is outside the control limits, namely the one for the $17th$ sample.

(b) Yes. The process is in control after the long run below the central line.

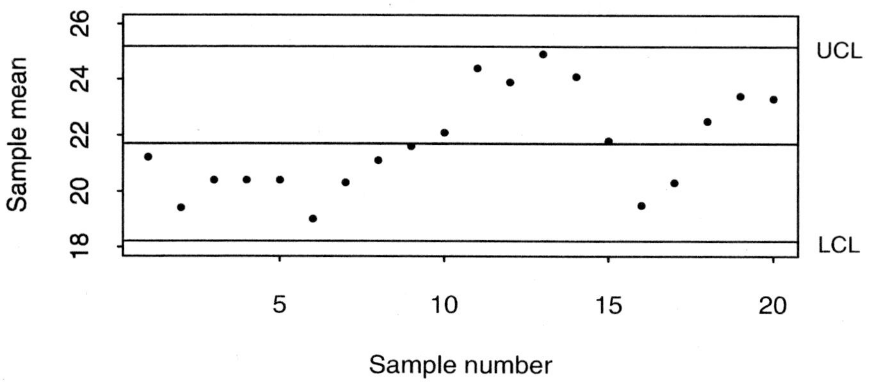

Figure 15.5: Control chart for sample mean. Exercise 15.5.

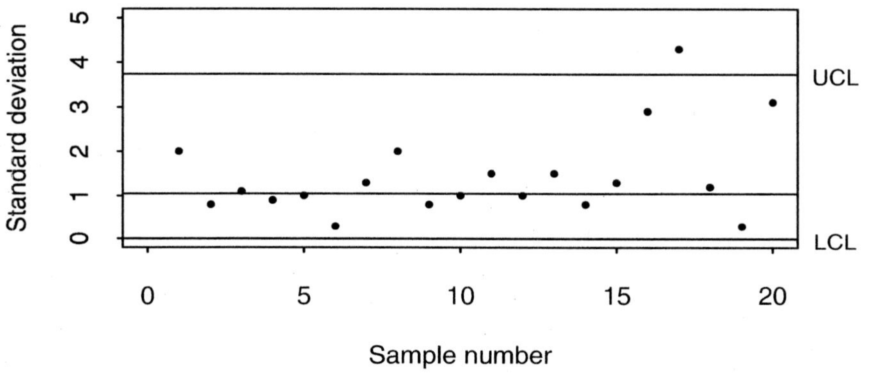

Figure 15.6: Control chart for sample standard deviation. Exercise 15.5.

15.7 The σ chart has central line $c_2\bar{s} = 0.8407\ (\ 0.002) = 0.00168$.

$$LCL = B_3\bar{s} = 0(0.002) = 0$$

$$UCL = B_4\bar{s} = 2.089(0.002) = 0.00418.$$

15.9 (a) For the data of Exercise 15.8, $\bar{p} = 0.0369$. Thus,

$$\begin{aligned} LCL &= \bar{p} - 3\sqrt{\bar{p}(1-\bar{p})/100} \\ &= 0.0369 - 3\sqrt{0.0369(1-0.0369)/100} = -0.020 \end{aligned}$$

which is taken to be 0, and

$$\begin{aligned} UCL &= \bar{p} + 3\sqrt{\bar{p}(1-\bar{p})/100} \\ &= 0.0369 + 3\sqrt{0.0369(1-0.0369)/100} = 0.0935 \end{aligned}$$

(b) The control chart is given in Figure 15.7. The process is in control.

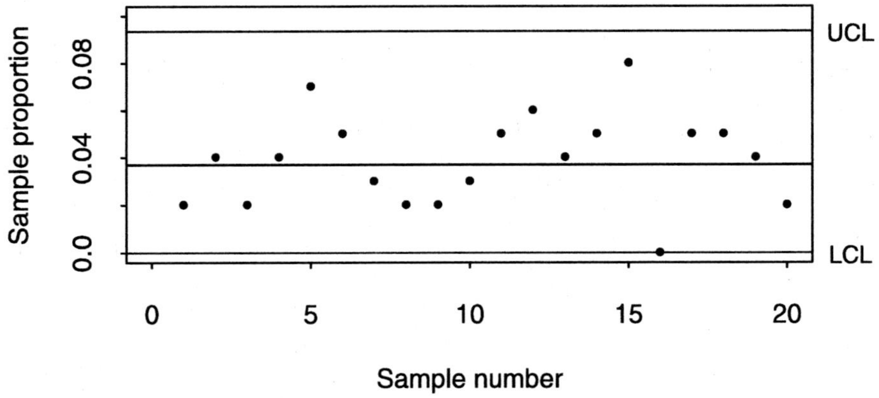

Figure 15.7: Control chart for the proportion defective for Exercise 15.9.

15.11 The central line is $\bar{c} = 4.92$. The LCL is

$$LCL = \bar{c} - 3\sqrt{\bar{c}} \;=\; -0.173$$

so the LCL is taken to be 0. The UCL is

$$UCL = \bar{c} + 3\sqrt{\bar{c}} \;=\; 11.57$$

The control chart is given in Figure 15.8. The process is in control.

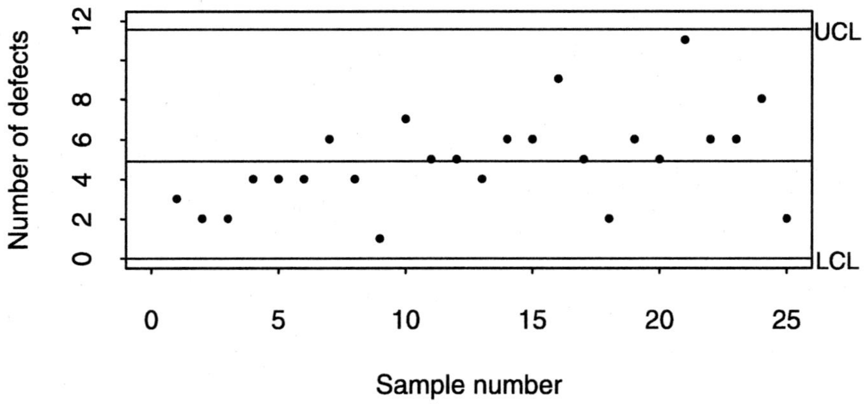

Figure 15.8: Control chart for number of defects for Exercise 15.11.

15.13 From Table 10, the tolerance limits are

$$\bar{x} \pm Ks = 52,800 \pm 3.457(4600)$$

or, from 36,897.8 to 68,702.2 . This means that with 95 percent confidence, 99 percent of the pieces will have a yield stress between 36,897.8 and 68,702.2 psi.

15.15 (a) $\bar{x}$ =0.1063, s = 0.0040, K = 2.052. Thus, the tolerance limits are

$$\bar{x} \pm Ks = 0.1063 \pm 2.052(0.0004)$$

or, from 0.1054 to 0.1072 .

(b) The confidence interval is given by

$$\bar{x} - t_{\alpha/2}\frac{s}{\sqrt{n}} < \mu < \bar{x} + t_{\alpha/2}\frac{s}{\sqrt{n}}$$

Since $n = 40$, the student's t quantile $t_{\alpha/2} = t_{.025}$ is well approximated by the standard normal quantile $z_{.025} = 1.96$. The 95 percent confidence interval is

$$.1063 - 1.96\frac{.0004}{\sqrt{40}} < \mu < .1063 + 1.96\frac{.0004}{\sqrt{40}}$$

or $.1062 < \mu < .1064$. For the difference between tolerance limits and confidence intervals, see the discussion in Section 14.4 of the text.

15.17 (a) The central line for the $\bar{x}$ chart is given by $y = \mu = 4.1$. The lower control limit is given by

$$LCL = 4.1 - \frac{3}{\sqrt{5}}(0.05) = 4.013$$

and the upper control limit is given by

$$UCL = 4.1 + \frac{3}{\sqrt{5}}(0.05) = 4.187.$$

(b) The central line for the R chart is given by $y = d_2\sigma = 2.326(0.05) = 0.163$. The lower control limit is given by $LCL = D_1\sigma = 0(0.05) = 0$ and the upper control limit is given by $UCL = D_2\sigma = 4.918(0.05) = 0.246$.

(c) The control charts are given in the Figures 15.9 and 15.10. For the $\bar{x}$ chart, points at 1, 3, 4, 5, 7, 8, 9, and 11 to 20 are outside the limits. For the R chart, the point at 6 is the only one outside of the limits.

15.19 If p is 0.02, then the central line is at 0.02, the

$$LCL = p - 3\sqrt{p(1-p)/200} = 0.02 - 3\sqrt{0.02(1-0.02)/200} = -0.0097$$

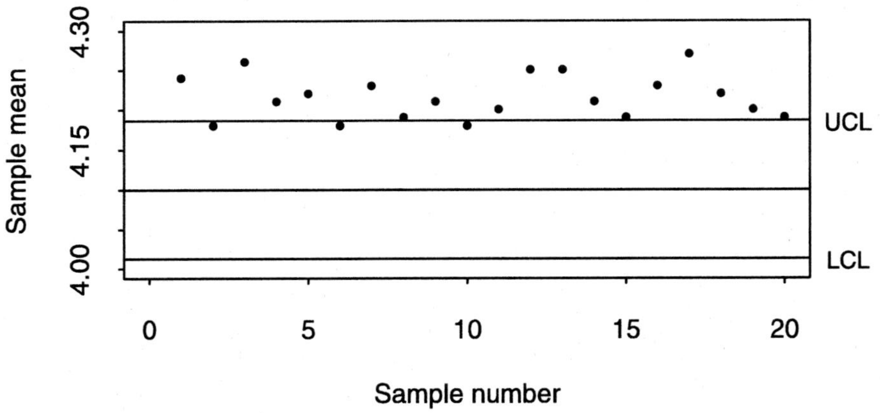

Figure 15.9: Control chart for the sample means for Exercise 15.17.

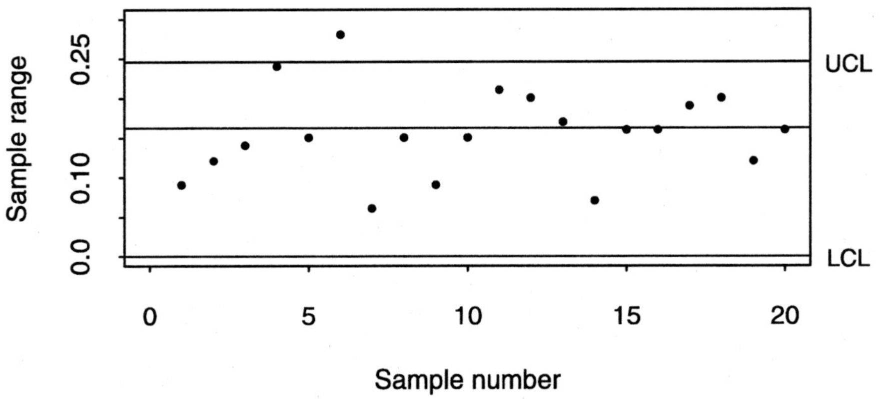

Figure 15.10: Control chart for the sample ranges for Exercise 15.17.

which is taken to be 0, and

$$UCL = p + 3\sqrt{p(1-p)/200} \ = \ 0.02 + 3\sqrt{0.02(1-0.02)/200} \ = \ 0.0497$$

The control chart is given in Figure 15.11. There are five points that are outside the control limits, so the standard is not being met.

15.21 The central line is $\bar{c} = 0.8$. The LCL is

$$LCL = \bar{c} - 3\sqrt{\bar{c}} \ = \ -1.88$$

so the LCL is taken to be 0. The UCL is

$$UCL = \bar{c} + 3\sqrt{\bar{c}} \ = \ 3.48$$

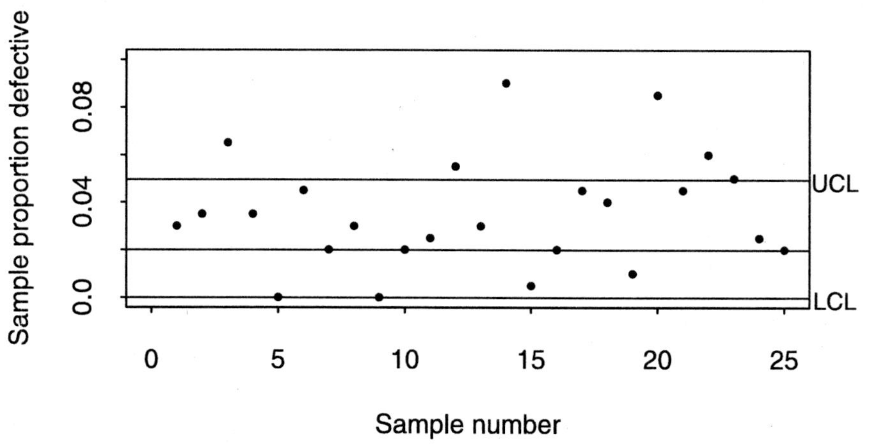

Figure 15.11: Control chart for the fraction defective for Exercise 15.19.

The control chart is given in Figure 15.12. Only the point at 19 is outside of the limits.

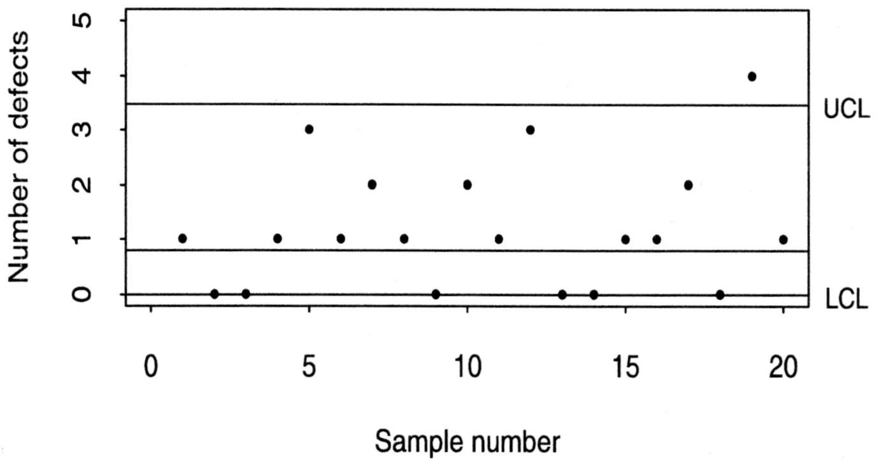

Figure 15.12: Control chart for number of defects for Exercise 15.21.

15.23 The $n = 50$ transformed observations $y = \ln x$ have $\bar{y} = 8.846$ and $s_y = 1.0293$ From Table 10 $K = 1.996$ so the 95 percent tolerance limits on proportion $P = .90$ for the transformed observations are

$$\bar{x} \pm Ks = 8.846 \pm 1.996(1.2093)$$

or 6.7915 to 10.9005. Converting these to the original scale, the tolerance limits are exp(6.7915) to exp(10.9005) or 890 to 54,203. This means that, with 95 percent confidence, 90 percent of the inter-request times will be between 890 and 54,203.

15.25 (a) We check that $\bar{x}$ =237.32, $s = 25.10$, $K = 2.126$. Thus, the tolerance bound is

$$\bar{x} - Ks = 237.32 - 2.126(25.10)$$

or 183.

(b) Here $K = 2.010$. Thus, the tolerance bound is

$$\bar{x} - Ks = 237.32 - 2.010(25.10)$$

or 186.

(c) The normal-scores plot reveals no marked departures from the assumption of normal observations.

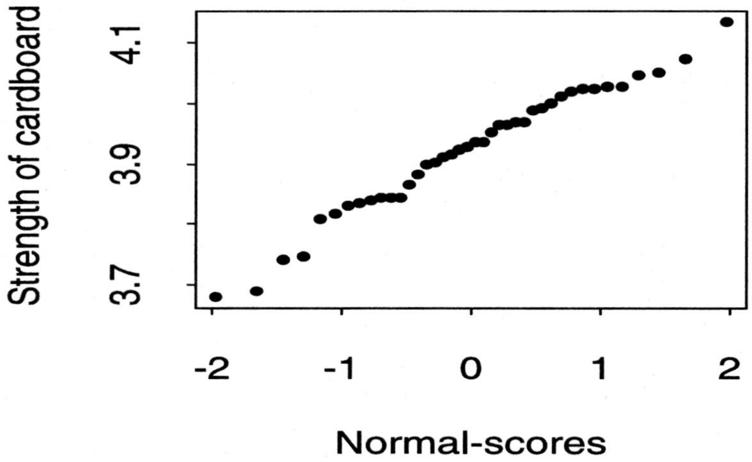

15.27 The sample mean and sample standard deviation are

$$\bar{x} = 2.5064 \quad \text{and} \quad s = .0022.$$

(a)

$$\hat{C}_p = \frac{USL - LSL}{6s} = \frac{2.516 - 2.496}{6(.0022)} = 1.515.$$

(b)

$$\begin{aligned}\hat{C}_{pk} &= \frac{\min(\bar{x} - LSL,\ USL - \bar{x})}{3s} \\ &= \frac{\min(2.5064 - 2.4960,\ 2.5160 - 2.5064)}{3(.0022)} \\ &= \frac{\min(.0104,\ .0096)}{3(.0022)} = 1.45\end{aligned}$$

Chapter 16

APPLICATION TO RELIABILITY AND LIFE TESTING

16.1 Since this is a series system , we need to find R such that

$$R^8 = .95 \quad \text{so} \quad R = (.95)^{1/8} = .9936.$$

16.3 The reliability of the B, C parallel segment is :

$$R_{BC} = 1 - (1 - .80)(1 - .90) = .980.$$

The reliability of the E, F parallel segment is :

$$R_{EF} = 1 - (1 - .90)(1 - .85) = .985.$$

The reliability of the A, B, C combination segment is :

$$R_{ABC} = R_A \; R_{BC} = (.95)(.980) = .931.$$

The reliability of the D, E, F combination segment is :

$$R_{DEF} = R_D \; R_{EF} = (.99)(.985) = .97515.$$

Thus, the reliability of the system is:

$$R = 1 - (1 - R_{ABC})(1 - R_{DEF})$$

$$= 1 - (1 - .931)(1 - .97515) = .9983.$$

16.5 (a) We have that

$$f(t) = Z(t)\, exp\,[\, -\int_0^t Z(t)\, dt\,].$$

where

$$Z(t) = \begin{cases} \beta(1 - t/\alpha) & \text{for } 0 < t < \alpha. \\ 0 & \text{elsewhere.} \end{cases}$$

Thus

$$f(t) = \begin{cases} \beta(1 - t/\alpha)\, exp\,[\, -\int_0^t \beta(1 - x/\alpha)\, dx\,] & \text{for } 0 < t < \alpha. \\ 0 & \text{elsewhere.} \end{cases}$$

so

$$f(t) = \begin{cases} \beta(1 - t/\alpha)\, exp\,[\, -\beta(t - t^2/(2\alpha)\,)\,] & \text{for } 0 < t < \alpha. \\ 0 & \text{elsewhere.} \end{cases}$$

The distribution function is

$$F(t) = \begin{cases} \int_0^t f(x)\, dx & \text{for } 0 < t < \alpha. \\ 1 & \text{for } t > \alpha. \end{cases}$$

or

$$F(t) = \begin{cases} 1 - exp\,[\, -\beta(t - t^2/(2\alpha)\,)\,] & \text{for } 0 < t < \alpha. \\ 1 & \text{for } t > \alpha. \end{cases}$$

Note that $F(x)$ must have a jump at $t = \alpha$.

(b) Since $\int_0^\alpha f(x)\, dx = 1 - e^{-\beta\alpha/2}$ the probability of initial failure is $= 1 - e^{-\alpha\beta/2}$

16.7 (a) Since the failure rate is constant,

$$f(t) = (.02)e^{-(.02)t}$$

and

$$F(t) = 1 - e^{-(.02)t}.$$

The unit of time is 1000 hours. The probability that the chip will last longer than 20,000 hours is

$$1 - F(20) = e^{-(.02)\cdot 20} = .6703.$$

(b) The 5000-hour reliability of four such chips in a series is

$$R = [1 - F(5)]^4 = e^{-4(.02)\cdot 5} = .6703.$$

16.9 The Weibull density is

$$f(t) = \alpha\beta t^{\beta-1}e^{-\alpha t^\beta} \quad \text{for} \quad \alpha, \beta, t > 0.$$

Thus,

$$F(t) = 1 - e^{-\alpha t^\beta}.$$

The probability that the component will operate at least 5,000 hours is

$$1 - F(5000) = e^{-(.005)(5000)^{.80}} = .0106 \quad \text{for} \quad \alpha, \beta, t > 0.$$

16.11 (a)

$$\begin{aligned} T_r &= \sum_{i=1}^{r} t_i + (n-r)t_r \\ &= 250 + 380 + 610 + 980 + 1250 + 30 \cdot 1250 = 40,970. \end{aligned}$$

Since $\chi^2_{.995}$ with 10 degrees of freedom is 2.156 and $\chi^2_{.005}$ with 10 degrees of freedom is 25.118, the 99 percent confidence interval is

$$\frac{2 \cdot 40,970}{25.118} < \mu < \frac{2 \cdot 40,970}{2.156}$$

or

$$3,253.1 < \mu < 38,005.6$$

(b) The null hypothesis is $\mu = 5,000$ and the alternative is $\mu > 5,000$. Since $\chi^2_{.05}$ with $2 \cdot r = 10$ degrees of freedom is 18.307, we reject the null hypothesis at the .05 level if

$$T_r > \frac{1}{2}\mu_0\chi^2_{.05} = \frac{1}{2}(5,000)(18.307) = 45,767.5$$

Since $T_r = 40{,}970$, we cannot reject the null hypothesis at the .05 level of significance. We cannot be sure the manufacturer's claim is true.

16.13 (a)

$$\begin{aligned} T_r &= \sum_{i=1}^{r} t_i + (n-r)t_r \\ &= 211 + 350 + 384 + 510 + 539 + 620 + 715 = 3,329. \end{aligned}$$

Since $\chi^2_{.975}$ with 14 degrees of freedom is 5.629 and $\chi^2_{.025}$ with 14 degrees of freedom is 26.119, the 95 percent confidence interval is

$$\frac{2 \cdot 3329}{26.119} < \mu < \frac{2 \cdot 3329}{5.629}$$

or

$$254.9 < \mu < 1,182.8$$

(b) The null hypothesis is $\mu = 500$ and the alternative is $\mu \neq 500$. Since $\chi^2_{.05}$ with 14 degrees of freedom is 23.658 and $\chi^2_{.95}$ with 14 degrees of freedom is 6.571, we reject the null hypothesis at the .10 level if

$$T_r < \frac{1}{2}\mu_0\chi^2_{.95} = \frac{1}{2}(500)(6.571) = 1,642.75$$

or if

$$T_r > \frac{1}{2}\mu_0\chi^2_{.05} = \frac{1}{2}(500)(23.685) = 5,921.25$$

Since $T_r = 3{,}329$, we cannot reject the null hypothesis at the .10 level of significance.

16.15 Since $2T_r/\mu$ is a χ^2 random variable with $2r$ degrees of freedom ,

$$P[\ \chi^2_{1-\alpha/2} < \frac{2T_r}{\mu} \quad and \quad \frac{2T_r}{\mu} < \chi^2_{\alpha/2}\ ] = \alpha,$$

where $\chi^2_{1-\alpha/2}$ and $\chi^2_{\alpha/2}$ are the chi-square quantiles for $2r$ degrees of freedom. Multiplying the first inequality by $\mu/\chi^2_{1-\alpha/2}$ and the second by $\mu/\chi^2_{\alpha/2}$ gives:

$$P\ [\ \mu < \frac{2T_r}{\chi^2_{1-\alpha/2}} \quad \text{and} \quad \frac{2T_r}{\chi^2_{\alpha/2}} < \mu\ ] = \alpha.$$

Thus the $(1-\alpha)100$ percent confidence interval is:

$$\frac{2T_r}{\chi^2_{\alpha/2}} < \mu < \frac{2T_r}{\chi^2_{1-\alpha/2}}$$

16.17 There are $r = 3$ failures and total time on test is

$$T_3 = 2076 + 3667 + 9102 + 197(9102) = 1,807,939$$

Since $\chi^2_{.05} = 12.592$ for $2r = 6$ degrees of freedom, the 95 percent lower confidence bound is

$$\frac{2T_3}{\chi^2_{.05}} = \frac{2(1,807,939)}{12.592} = 287,156.8$$

16.19 The probability the diaphragm valve will perform at least 150 hours is 1 minus the probability that it fails before 150 hours. Since

$$1 - F(150) = 1 - (1 - e^{-\alpha(150)^\beta}),$$

$\hat{\alpha} = .0105$ and $\hat{\beta} = .5062$, we estimate this probability by

$$e^{-\hat{\alpha}(150)^{\hat{\beta}}} = e^{-(.0105)(150)^{.5062}} = .8758.$$

16.21 Suppose T has the Weibull distribution with parameters α and β. In the text it was shown that

$$E(T) = \frac{1}{\alpha^{1/\beta}}\Gamma(1+\frac{1}{\beta})$$

Since $Var(T) = E(T^2) - [E(T)]^2$, we must find $E(T^2)$. The Weibull density is

$$\alpha\beta t^{\beta-1}e^{-\alpha t^{\beta}}$$

so

$$E(T^2) = \int_0^{\infty} t^2\alpha\beta t^{\beta-1}e^{-\alpha t^{\beta}}\, dt.$$

Let $u = \alpha t^{\beta}$. Then

$$E(T^2) = \frac{1}{\alpha^{2/\beta}}\int_0^{\infty} u^{2/\beta}e^{-u}du = \frac{1}{\alpha^{2/\beta}}\Gamma(1+\frac{2}{\beta})$$

so that

$$\begin{aligned} Var(T) &= \frac{1}{\alpha^{2/\beta}}\Gamma(1+\frac{2}{\beta}) - (\frac{1}{\alpha^{1/\beta}}\Gamma(1+\frac{1}{\beta}))^2 \\ &= \frac{1}{\alpha^{2/\beta}}\cdot[\,\Gamma(1+\frac{2}{\beta}) - (\Gamma(1+\frac{1}{\beta}))^2\,] \end{aligned}$$

16.23 (a) The probability of failure during the first 250 hours of operation is:

$$1 - e^{-(.0045)\cdot 250} = .6753.$$

(b) The probability that two independent components will survive the first 100 hours of operation is:

$$e^{-(.0045)\cdot 100}\cdot e^{-(.0045)\cdot 100} = .4066.$$

16.25 There are $r = 4$ failures and total time on test is

$$T_4 = 3582 + 8482 + 8921 + 16303 + 296(16303) = 4,862,976$$

Since $\chi^2_{.05} = 15.507$ for $2r = 8$ degrees of freedom, the 95 percent lower confidence bound is

$$\frac{2T_4}{\chi^2_{.05}} = \frac{2(4,862,976)}{15.507} = 627,197.5$$

16.27 The probability the circuit will perform at least 100 hours is 1 minus the probability that it fails before 100 hours. Since

$$1 - F(100) = 1 - (1 - e^{-\alpha(100)^{\beta}})$$

$\hat{\alpha} = .0000909$ and $\hat{\beta} = 1.3665$, we estimate this probability by

$$e^{-\hat{\alpha}(100)^{\hat{\beta}}} = e^{-(.0000909)(100)^{1.3665}} = .9520.$$

16.29 (a) We are given that X has

$$F(x) = 1 - e^{-.01x} \quad and \quad f(x) = .01e^{-.01x}$$

and Y has distribution

$$G(y) = 1 - e^{-.005y} \quad and \quad g(y) = .005e^{-.005y}.$$

Consequently,

$$\begin{aligned}
R &= P[Y > X] = \int_{-\infty}^{\infty} F(y)g(y)\,dy \\
&= \int_{-\infty}^{\infty} [1 - e^{-.01y}].005e^{-.005y}\,dy \\
&= \int_{-\infty}^{\infty} .005e^{-.005y}dy - \int_{-\infty}^{\infty} (.005)e^{-(.005+.01)y}\,dy \\
&= 1 - \frac{.005}{.015} = .6667.
\end{aligned}$$

(b) We are given that X has

$$F(x) = 1 - e^{-.005x} \quad and \quad f(x) = .005e^{-.005x}$$

and Y has distribution

$$G(y) = 1 - e^{-.005y} \quad and \quad g(y) = .005e^{-.005y}.$$

Consequently,

$$\begin{aligned} R &= P[Y > X] = \int_{-\infty}^{\infty} F(y)g(y)\,dy \\ &= \int_{-\infty}^{\infty} [1 - e^{-.005y}](.005)e^{-.005y}\,dy \\ &= \int_{-\infty}^{\infty} (.005)e^{-.005y}\,dy - \int_{-\infty}^{\infty} .005e^{-(.005+.005)y}\,dy \\ &= 1 - \frac{.005}{.010} = .50 \end{aligned}$$

(c) Since $\ln X$ and $\ln Y$ are independent and each has a normal distribution, $\ln Y - \ln X$ has a normal distribution with mean $\mu_y - \mu_x = 80 - 60 = 20$ and variance $\sigma_x^2 + \sigma_y^2 = 5^2 + 5^2 = 50$. To avoid numerical integration, we note that

$$\begin{aligned} R &= P[Y > X] = P[\ \ln Y > \ln X] = P[\ \ln Y - \ln X > 0] \\ &= 1 - F(\frac{-20}{\sqrt{50}}) = 1 - F(-2.83) \end{aligned}$$

where $1 - F(-2.83) = F(2.83) = .9977$ is obtained from the standard normal distribution.